WOODLAND HABITAT

Trees and woodlands are an integral part of Britain's heritage and culture, as well as an invaluable environmental and ecological resource. In an increasingly urbanised British population, romantic notions of bluebell-carpeted woods and deep-coloured wooded hill-sides during autumn are a distant memory. Most peoples' experiences of woods are of forgotten corners of land with a few sycamore trees and lots of brambles, or an area of dark, intimidating conifers are more common. The wide variety of types of woodland, whether closely managed or naturally occurring, provide important habitats for a wide range of flora and fauna.

Woodland Habitats explores the history and ecology of British woodlands, and explains why they are such a valuable resource. It offers a practical guide to issues concerned with the ecology of woodland habitats and organisms; conservation and management; coppicing, pasture woodland and commercial forestry; woodland grazing, ride management and recreation in woodlands.

Featuring illustrated species boxes as well as a full species list, examples of notable sites with location maps and pictures, suggested projects and a full glossary, students and environmentalists will gain a deeper understanding of the historical and present-day importance of British woodlands, as both an ecological and a cultural resource.

Helen J. Read is site ecologist for Burnham Beeches, Buckinghamshire. **Mark Frater** is the managing director of a company specialising in woodland management.

HABITAT GUIDES
Series Editor: **C. Philip Wheater**

Other titles in the series:

Upland Habitats
Urban Habitats

Forthcoming titles:

Freshwater Habitats
Grassland and Heathland Habitats
Marine Habitats
Agricultural Habitats

WOODLAND HABITATS

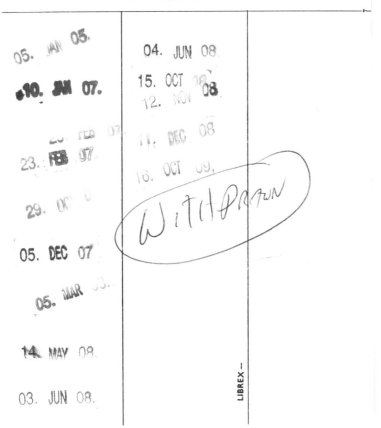

LONDON AND NEW YORK

First published 1999
by Routledge
11 New Fetter Lane, London EC4P 4EE

Simultaneously published in the USA and Canada
by Routledge
29 West 35th Street, New York, NY 10001

Transferred to Digital Printing 2002

Routledge is an imprint of the Taylor & Francis Group

© 1999 Helen J. Read and Mark Frater; Jo Wright for the illustrations

The right of Helen J. Read and Mark Frater to be identified as the Authors of this work has been
asserted by them in accordance with the Copyright, Designs and Patents Act 1988

Typeset in Sabon by RefineCatch Limited, Bungay, Suffolk
Printed and bound in Great Britain by
TJI Digital, Padstow, Cornwall

British Library Cataloguing in Publication Data
A catalogue record for this book is available from the British Library

Library of Congress Cataloging in Publication Data
Read, Helen J.
Woodland habitats / Helen J. Read & Mark Frater.
p.   cm. – (Habitat guides)
Includes bibliographical references (p.   ).
1. Forests and forestry – Great Britain.   2. Forest management –
Great Britain.   3. Forest conservation – Great Britain.   4. Forest
ecology – Great Britain.   I. Frater, Mark. 1953– .   II. Title.
III. Series.
SD179.R435   1999
333.75'15'0941 – dc21   98–45288

ISBN 0–415–18089–9 (hbk)
ISBN 0–415–18090–2 (pbk)

CONTENTS

•

PLATES

•

FIGURES

●

TABLES

●

SPECIES BOXES

•

ACKNOWLEDGEMENTS

•

We would like to thank all those who have helped with the production of this book especially Keith Kirby and Mike Martin whose comments improved the content of the book substantially. The editor (Phil Wheater) and the publishers have been very helpful thoughout the various stages and we would like to thank Phil for his comments on the text and for help in finding the authorities for some of the organisms mentioned. Roger Cook and Andy Swash also helped with the classification and locating authorities and Daniel Noble read through the script in the early stages. Any errors, however, are entirely attributable to us.

Several people took the time to show us round their various sites, provided information and then checked the accuracy of the case studies when written. We are grateful to Peter Fordham (Bradfield Woods), David Duncan (Inshriach), Graham Gill and Bill Burlton (Kielder), Doug Oliver (Coed y Rhygen), Andrew Barnard and Ian Turney (Burnham Beeches) for all their help.

We are also grateful to those who have lent photographs for use; Paul Read (Plates 1, 3 and 2.5), Ted Green (Plate 4) and Roger Key (Plate 5) and the Corporation of London (Plate 4.2) and to those who have given permission for the use of their figures and tables as listed below.

Figure 1.1 is from H. Godwin (1975), *History of the British Flora*, with permission from Cambridge University Press. Figures 1.2 and 1.3 are from P. Marren (1990), *Woodland Heritage*, David and Charles with permission from P. Marren. Figure 1.4 is from K.J. Kirby *et al.* (1984), *Inventories of Ancient Semi-natural Woodland*, with permission from English Nature. Figure 1.5 is from Forestry Commission/Countryside Commission (1996), *Woodland Creation: Needs and Opportunities in the English Countryside*, CCP 507, with permission from the Countryside Commission and the Forestry Commission. Figures 1.6 and 1.7 are from Forest Enterprise (1997), *Corporate Plan 1997–2000*, Forest Enterprise, Edinburgh, with permission from the Forestry Commission. Figure 2.1 is from H. Godwin (1975), *History of the British Flora*, with permission from Cambridge University Press. Figure 2.2 is from Kluwer Academic Publishers, *Functional Ecology of Woodlands and Forests*, by J.R. Packham, D.J.L. Harding, G.M. Hilton and R.A. Stuttard (1992), page 86, Figure 3.8, with kind permission from Kluwer Academic Publishers and Professor J. Packham. Figure 2.4 is from A.G. Tansley (1939), *The British Islands and their Vegetation*, with permission from Cambridge University Press. Figure 2.5 is from K. Heliövaara and R. Väisänen (1984), 'Effects of modern forestry on northwestern European forest invertebrates: a synthesis', *Acta Forestalia Fennica*, with permission of the publishers. Figure 2.6 is from K.J. Kirby and J.R. Heap (1984) 'Forestry and nature conservation in Romania', *Quarterly Journal of Forestry*, 78,

145–55 with kind permission of the Royal Forestry Society. Figure 2.7 is after B. Clouston and K. Stansfield (1979) *After the Elm*, published by William Heinemann with permission from Random House. Figure 2.8 is from B. Clouston and K. Stansfield (1979) *After the Elm*, published by William Heinemann with permission from Random House. Figure 3.1 is from R.J. Fuller and M.S. Warren (1993), *Coppiced Woodlands: Their Management for Wildlife* with permission from JNCC and *Woodland Conservation and Management* (1981) by G. Peterken, with kind permission from Kluwer Academic Publishers and G. Peterken. Figure 3.2a is from R.J. Fuller (1995), *Bird Life of Woodland and Forest*, with permission from Cambridge University Press. Figure 3.2b is from R.J. Fuller (1992), 'Effects of coppice management on woodland breeding birds', in Kluwer Academic Publishers, *Ecology and Management of Coppice Woodlands*, edited by G.P. Buckley, page 175, Figure 9.3 with kind permission from Kluwer Academic Publishers and Dr R. Fuller. Figure 3.3 is from M.S. Warren and J.A. Thomas (1992), 'Butterfly responses to coppicing', in Kluwer Academic Publishers, *Ecology and Management of Coppice Woodlands*, edited by G.P. Buckley, page 253, Figure 13.3 with kind permission from Kluwer Academic Publishers and M. Warren. Figure 3.4 is from D. Steel and N. Mills (1988), 'A study of the plants and invertebrates in an actively coppiced woodland (Brasenose Wood, Oxfordshire)', in *Woodland Conservation and Research in the Clay Vale of Oxfordshire and Buckinghamshire*, edited by K.J. Kirby and F.J. Wright, with permission from English Nature. Figure 3.6 is from C. Bowden and R. Hoblyn (1990), 'The increasing importance of restocked conifer plantations for woodlarks in Britain: implications and consequences', *RSPB Conservation Review*, with permission of the RSPB. Figure 3.8 is from C. Steel (1988), 'Butterfly monitoring in Sheephouse Wood', in *Woodland Conservation and Research in the Clay Vale of Oxfordshire and Buckinghamshire*, edited by K.J. Kirby and F.J. Wright, with permission from English Nature. Figure 3.9 is from C.R. Tubbs (1986), *The New Forest: A Natural History*, with permission from HarperCollins Publishers Ltd. Figure 3.10 is from Forestry Commission/Countryside Commission (1996), *Woodland Creation: Needs and Opportunities in the English Countryside*, CCP 507, with permission from the Countryside Commission and the Forestry Commission. Figure 4.2 is from H.J. Read *et al.* (1996), in *Pollard and Veteran Tree Management*, edited by H.J. Read with permission from the Corporation of London. Figure 4.3 is from P. Archer, P. Fordham, and M. Harding, (1995), 'Bradfield Woods: Management Plan', Suffolk Wildlife Trust, with permission. Figure 4.4 is from RSPB (1993) *Time for Pine* with kind permission of the RSPB. Figure 5.4 is from R.R. Askew and M. Redfern (1992), *Plant Galls*, with permission from R. Askew. Parts of Tables 2.8 and 2.9 are from *Deer Management and Woodland Conservation in England* (1997) with permission from English Nature. Figures 2.3, 2.7, 3.7 and those for the species boxes were all drawn by Jo Wright.

SERIES INTRODUCTION

•

The British landscape is semi-natural at best, having been influenced by human activities since the Mesolithic era (*circa* 10,000–4,500 BC). Although these influences are most obvious in urban, agricultural and forestry sites, there has been a major impact on those areas we consider to be our most natural. For example, upland moorland in northern England was covered by wild woodland during Mesolithic times, and at least some was cleared before the Bronze Age (*circa* 2,000–500 BC), possibly to extend pasture land. The remnants of primeval forests surviving today have been heavily influenced by their usage over the centuries, and subsequent management as wood-pasture and coppice. Even unimproved grassland has been grazed for hundreds of years by rabbits introduced, probably deliberately by the Normans, sometime during the twelfth century.

More recent human activity has resulted in the loss of huge areas of a wide range of habitats. Recent government statistics record a 20 per cent reduction in moorland and a 40 per cent loss of unimproved grassland between 1940 and 1970 (Brown, 1992). In the forty years before 1990 we lost 95 per cent of flower-rich meadows, 60 per cent of lowland heath, 50 per cent of lowland fens and ancient woodland, and our annual loss of hedgerows is about 7,000 km. There has been substantial infilling of ponds, increased levels of afforestation and freshwater pollution, and associated reductions in the populations of some species, especially rarer ones. These losses result from various impacts: habitat removal due to urban, industrial, agricultural or forestry development; extreme damage such as pollution, fire, drainage and erosion (some or all of which are due to human activities); and other types of disturbance which, although less extreme, may still eradicate vulnerable communities. All of these impacts are associated with localised extinctions of some species, and lead to the development of very different communities to those originally present. During the twentieth century over one hundred species are thought to have become extinct in Britain, including 7 per cent of dragonfly species, 5 per cent of butterfly species and 2 per cent of mammal and fish species. Knowledge of the habitats present in Britain helps us to put these impacts into context and provide a basis for conservation and management.

A habitat is a locality inhabited by living organisms. Habitats are characterised by their physical and biological properties, providing conditions and resources which enable organisms to survive, grow and reproduce. This series of guides covers the range of habitats in Britain, giving an overview of the extent, ecology, fauna, flora, conservation and management issues of specific habitat types. We separate British habitats into seven major types and many more minor divisions. However, do not be misled into thinking that the natural world is easy to place into pigeon holes. Although these are convenient divisions, it is important to recognise that there is considerable commonality between the major habitat types which form the basis of the volumes in this series. Alkali waste tips in urban areas provide similar conditions to calcareous grasslands, lowland heathland requires similar management regimes to some heather moorland, and both estuarine and lake habitats may suffer from similar problems of accretion of sediment. In contrast, within each of the habitat types discussed in individual volumes, there may be great differences: rocky and sandy shores, deciduous and coniferous woodlands, calcareous and acid grasslands are all typified by different plants and animals exposed to different environmental conditions. It is important not to become restricted in our appreciation of the similarities which exist between apparently very different habitat types and the, often great, differences between superficially similar habitats.

The series covers the whole of Britain, a large geographical range across which plant and animal communities differ, from north to south and east to west. The climate, especially in temperature range and precipitation, varies throughout Britain. The south-east tends to experience a continental type of climate with a large annual temperature range and maximal rainfall in the summer months. The west is influenced by the sea and has a more oceanic climate, with a small annual temperature range and precipitation linked to cyclonic activity. Mean annual rainfall tends to increase both from south to north and with increased elevation. Increased altitude and latitude are associated with a decrease in the length of the growing season. Such climatic variations support different species to differ-

ing extents. For example, the small-leaved lime, a species which is thermophilic (adapted for high temperatures), is found mainly in the south and east, while the cloudberry, which requires lower temperatures, is most frequent on high moorland in the north of England and Scotland. Equivalent situations occur in animals. It is, therefore, not surprising that habitats of the same basic type (such as woodland) will differ in their composition depending upon their geographical location.

In the series we aim to provide a comprehensive approach to the examination of British habitats, while increasing the accessibility of such information to those who are interested in a subset of the British fauna and flora. Although the series comprises volumes covering seven broad habitat types, each text is self-contained. However, we remind the reader that the plants and animals discussed in each volume are not unique to, or even necessarily dominant in, the particular habitats but are used to illustrate important features of the habitat under consideration. The use of scientific names for organisms reduces the likelihood of confusing one species with another. However, because several groups (especially birds, and to a lesser extent, flowering plants) are often referred to by common names (and for brevity), we use common names where possible in the text. We have tried to use standard names, following a recent authority for each taxonomic group (see the species list for further details). Where common names are not available (or are confusing), the scientific name has been used. In all cases, species mentioned in the text are listed in alphabetical order in the species index and, together with the scientific name, in systematic order in the species list.

1

INTRODUCTION

•

GENERAL

Trees and woodlands are an integral part of Britain's heritage and culture. From pictures of kings hunting in leafy forests to the works of Thomas Hardy with their detailed descriptions of the lives of woodland occupants they are a part of our landscape which is often taken for granted (Plate 1.1). This is perhaps not surprising; our ancestors relied on wood to keep them warm, to provide fuel for cooking, as timber for building houses and ships and to make a myriad of other products which today we are no longer familiar with. Even now, the products of trees are central to our lives. The quantity of paper and packaging everyone

uses is substantial and, although houses are no longer built solely from wood, it is still widely used in construction as well as in furniture and fittings.

Romantic notions of oak woods full of bluebells in the spring or the glorious colours of an autumn beech wood are, for many people, just a distant memory. As the population becomes increasingly urbanised, most people's experiences of woods are scrappy forgotten corners of land with a few sycamore trees and lots of brambles, or, a large, seemingly intimidating, area of dark conifers. Research has shown that many visitors dislike areas of dense woodland because they *feel* unsafe (Countryside Commission, 1995). Yet, as the

Plate 1.1: A woodland church in Exmoor

population is more mobile and millions flock to areas such as the Lake District for holidays, parts of our landscape including our woods, are becoming holiday resorts.

Although the British woods have been shaped, and often formed, by human hand, this is not to say that they are important just for their products. The extensive felling of trees by humans after the last ice age substantially reduced the total area of woodland. The areas left, especially those that have never been clear felled, represent important resources for naturally occurring plants and animals. The complex three-dimensional structure of woods abounds with conditions suitable for a wide range of organisms. Some are generalists feeding on the dead leaves from the trees when they fall, others are incredibly specialised, living, for example, in a specific type of rot found on a particular tree species. As centres of diversity, woodlands are extremely important.

Woods and forests protect soils, retain moisture and store and recycle nutrients. They are also an essential part of the global environment, absorbing carbon dioxide and producing oxygen. They play a major role in maintaining the world climate, and their destruction releases carbon into the atmosphere causing climate change. More than 40 per cent of the earth's land surface used to be covered in trees. Today that has been reduced by a third and half of that remaining is degraded from its primary form (Myers, 1996).

This volume will explore the history and ecology of British woodlands and provide a description of the resource we have today. The management and conservation of woods is a complex subject which has stimulated much debate recently and which will, no doubt, continue well into the future. Case studies are presented of specific sites, which are good examples of particular types or illustrate relevant issues. Finally some examples of practical projects are given with the aim of stimulating the reader's interest in trees and woods wherever they live.

THE HISTORY OF WOODLAND IN BRITAIN

The last ice age to 4,500 years ago

The Pleistocene era in Britain was marked by a series of glacial and interglacial periods. The last ice age is known as the Weichselian and commenced about 70,000 years ago. Although the ice sheet only reached as far south as East Anglia the conditions further south in the British Isles would have been similar to tundra. As the temperature warmed up and the ice retreated (roughly 10,000 years ago), trees slowly colonised from further south. At this stage Britain was not an island so plants and animals could spread from the rest of Europe without having to cross any sea.

The progression of the trees and the species concerned can be determined by the use of pollen analysis. The pollen grains from different plant species or genera can be distinguished. When the grains fall onto the surface of a peat bog, or wet area where peat is forming, they are preserved. Over thousands of years, as the peat builds up so the pollen grains of the trees, and other plants common at that time, also accumulate and a record is formed. This can be examined by extracting a core of the peat and examining it under a microscope.

Figure 1.1 shows the colonisation of the British Isles by different tree species. Around 4,500 years ago most of the country was probably wooded and Figure 1.2 shows the likely major composition of the woods in

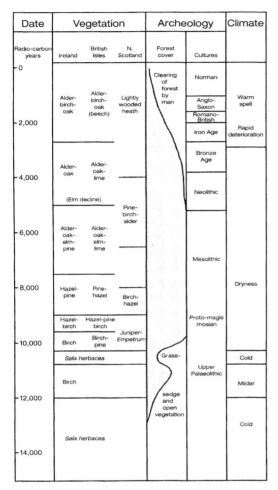

Date	Vegetation			Archeology		Climate
Radio-carbon years	Ireland	British Isles	N. Scotland	Forest cover	Cultures	
0		British Isles		Clearing of forest by man	Norman	Warm spell
	Alder-birch-oak	Alder-birch-oak (beech)	Lightly wooded heath		Anglo-Saxon	
2,000					Romano-British	Rapid deterioration
					Iron Age	
	Alder-oak	Alder-oak-lime			Bronze Age	
4,000					Neolithic	
	(Elm decline)		Pine-birch-alder			
6,000	Alder-oak-elm-pine	Alder-oak-elm-lime			Mesolithic	
						Dryness
8,000	Hazel-pine	Pine-hazel	Birch-hazel			
	Hazel-birch	Hazel-pine birch			Proto-magle mosian	
	Birch	Birch-pine	Juniper-Empetrum			
10,000						
	Salix herbacea			Grass-	Upper Palaeolithic	Cold
	Birch					Milder
12,000				sedge and open vegetation		Cold
	Salix herbacea					
14,000						

Figure 1.1 The vegetation of the British Isles in the last 14,000 years. From Godwin (1975).

different parts of the country. The woodland of this time is called wildwood. This was a period of relative climatic stability (called the Atlantic period) which ended with a decline in the amount of elm pollen recorded and an increase in human activities. These two events may be linked, but other causes of the elm decline have been proposed including Dutch elm disease (Rackham, 1990).

From this point onwards humans start to have a significant impact on the woodlands in the British Isles.

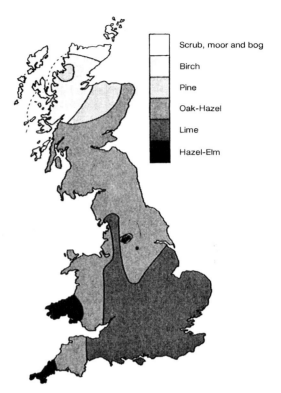

Scrub, moor and bog

Birch

Pine

Oak-Hazel

Lime

Hazel-Elm

Figure 1.2 The principal woodland types of the 'wildwood' (in addition there would have been alder dominated woods alongside lakes and rivers). From Marren (1990).

4,500 years ago to 1914 AD

Woodland clearance started in the Neolithic period and continued thoughout the Bronze Age (approximately 4000–2500 BP). Rackham (1990) considers that the period of greatest clearance was perhaps the Iron Age (approximately 2500–2000 BP) by the end of which 50 per cent of England had ceased to be wooded. No mean feat for people with primitive tools! Due to human activities the tree composition also changed. The amount of small-leaved lime decreased while hazel (and to a lesser extent birch) became more abundant (Godwin, 1975; Rackham, 1990). The forests were not all destroyed, some were probably used for grazing and fodder for domesticated animals, while the trees themselves started to become a managed resource.

From Neolithic days humans have used the resulting growth when a tree was felled and the stump re-grew (coppice) for putting on trackways, some being interwoven like hurdles (Rackham, 1990). Thus the management of the woods and the extensive use of woodland products started to become an integral part of British life.

From 2,800 years ago up to the present day it becomes difficult to disentangle small-scale changes in climate from the impact of humans in the pollen record (Godwin, 1975). Though it seems difficult to be sure of the exact timing of the various events concerning woodlands, many characteristic features such as the banks surrounding them may date from before the Anglo-Saxon period (Rackham, 1990). At the time of the Domesday Book in 1086 only 15 per cent of England was wooded (Rackham, 1980). By 1200 AD Rackham (1990) considers that the British woods had their boundaries and ownership defined, had acquired names and that woodland management by coppice-type methods was widespread. Woodland clearance continued thoughout the medieval period with woodland being destroyed at a rate of at least 20 acres per day (Rackham, 1990).

From the thirteenth century onwards more detailed records start to become available for individual woods, describing them, their management and financial recompense for the products derived from them. Rackham (1990) gives a detailed description of this period with examples and he also notes that the history of Wales, Scotland and Ireland differs from that of England. Much of the woodland management we now regard as 'traditional' was probably taking place during this time. Timbers, usually of oak, were used for the frames of houses and the underwood, or products of the

coppice, were used for a multitude of purposes from the walls of houses to the handles of tools. The village people often had rights of common such as pasturage, pannage (turning out pigs in the autumn months) and estovers (cutting wood for fuel and/or collecting fallen wood). The land, and the timber trees or the trunks of the trees, were usually owned by the Lord of the Manor.

In the period from 1066 onwards, the parks and Royal Forests were established. Parks were areas of private land surrounded by a fence, known as a pale, in which the owner usually kept deer (Rackham, 1988). Royal Forests were large areas of land which were set aside by the King for hunting (and preserving) deer for which he could create byelaws to protect. The forest areas were not fenced and the deer were free to wander. The Royal Forests were by no means just woodland. Some like Dartmoor and Exmoor had little woodland, others, like the New Forest had areas of woodland interspersed with heathland and grassland. If the Forest included former common land, the local people continued to have their common rights but the deer belonged to the King and none could be taken without his permission. In 1300 a total of about 200,000 hectares were part of Royal Forests (Marren, 1990), distributed as shown in Figure 1.3. However, this amount is a decline from that of 1200. In 1300 there were also about 3,200 parks. While the formation of the Forests did not prevent all of the woodland within them being cleared, it did slow the rate of clearance.

As the importance of the Royal Forests for hunting started to decline, the need for woodland for timber (large pieces of wood) increased. This demand became increasingly important as England developed a navy. The wooden ship building period, from 1750 to 1860, made extensive use of big timber taken from large mature trees. Ships were mainly

Figure 1.3 The distribution of Royal Forests in 1300. From Marren (1990).

made of oak (Species Box 1.1) so it was the oak woods which were most affected by the demand. At much the same time there was an increased need for oak from the leather tanning industry which used the bark from smaller pieces of wood (often the product of oak coppice). While these industries may not have caused a decline in oak woodlands (Rackham, 1990), they certainly did have a significant impact on the composition of certain woods, an example being the New Forest where large quantities of oak were felled which then allowed a pulse of tree regeneration due to the increased light levels (Tubbs, 1986). Many of

today's older broad-leaved plantations were planted during this period (Aldhous, 1997).

By the beginning of the twentieth century coppicing, pollarding and other traditional management methods were declining. Coal and coke became more widely available as sources of fuel instead of wood and charcoal and industrial products started to replace wooden ones. Demand for home-grown timber, however, increased substantially again at the time of the First World War.

1914 to the present day

Between 1914 and 1918 timber was felled on between 180,000 and 200,000 hectares of private woodland (Aldhous, 1997). As a result of this, and the low profit margin made by woodland owners just prior to the war, the Government formed the Forestry Commission in 1919. Their principal target was to create a strategic timber reserve in the form of a total of 2 million hectares of productive forestry by the end of the century. This was achieved by 1996 (Aldhous, 1997). (Note that forests in this context are *commercial forests*, which are not at all the same as the Royal Forests of medieval times.)

The fellings for the First World War reduced the area of woodland with coniferous trees on it to its lowest ever level (Aldhous, 1997). Broadleaves were also extensively felled, having a significant impact on some woods (Hornby, 1988). The new plantings, as a consequence of the wartime experiences and overall demand, were largely coniferous. Fast-growing exotic species were imported from other parts of the world (especially North America) and after some trials, building on previous experiences, these were widely planted in place of British native species.

Table 1.1 shows the amount of woodland in

....

Species Box 1.1: Oak

There are two species of oak native to Britain, pedunculate and sessile. The leaves of both are characteristically lobed but differ in the shape of the lobes, the length of stalk and the presence of auricles or folds of the leaf blade close to the stalk. For most of the year the leaves are usually a dull matt green colour, but more noticeable are the young shoots produced later in the summer which can be bright green or reddish in colour and are often called lammas growth. Lammas growth and regrowth from cut stumps seems to be particularly badly affected by a powdery mildew fungus which turns the leaves silvery-white. The male flowers are on rather insignificant green tassels, while the female ones are pink and tiny at the ends of the shoots. The fruit of the oak is the acorn, which sits in a cup either directly onto the twig (sessile oak) or on a stalk (pedunculate oak). The wood of both trees is very hard and strong and is used for a very wide range of products, construction timber being the most important. The light grey bark tends to be rough and strongly corrugated in mature and older trees. Traditionally it was used for tanning leather. Both species have a long tap root but they tend to be found in different conditions. The pedunculate oak occurs in deeper, moister soils whereas the sessile oak is more characteristic of poorer, well-drained areas. Oak trees are very long-lived and as they reach old age tend to shed branches and die back from the top so reducing the size of the crown. Trees over 400 years are relatively common.

Source: White (1995)

....

Britain at various points in time. The tremendous increase in the area of productive coniferous woodland postwar can be seen. The decline in coppice is also worth noting.

The extensive planting concentrated on poor land which was not suitable for agricultural cultivation. As a consequence large areas of uplands in the north of England, and

Table 1.1: The area of woodland in Britain through time

(a) Broad-scale changes

Year	Woodland cover %
5000 BC	80
1000 AD	20
1086 AD	15
1300 AD	10
1900 AD	5
1997 AD	10

Source: Forestry Commission/Countryside Commission (1996)

(b) Detailed changes between 1895 and 1996 (areas in 000's hectares)

	Effective date of survey or census							
	1895	*1913*	*1924*	*1939*	*1947*	*1965*	*1980*	*1996*
Productive high forest	–	877	573	771	757	1,265	1,881	2,157
Conifer	–	522	328	464	403	915	1,321	1,521
Hardwood	–	355	245	307	354	350	560	636
Coppice/coppice with standards	–	246	215	220	146	30	39	40
Other woodland	–	171	466	385	554	445	188	209
Scrub	–	–	134	–	201	353	148	–
Felled	–	–	193	–	328*	–	40	–
Other	–	–	139	–	25	92	–	–
Total woodland	1,103	1,294	1,254	1,376	1,457	1,740	2,108	2,406

Source: Aldous (1997)

Note:
* The area of wartime felling was 200,000 hectares

in Wales and Scotland were planted up with trees. This did not meet with universal approval and objections to the blankets of foreign conifers covering the moorland were increasingly expressed. By the 1960s a landscape consultant had been appointed by the Forestry Commission to advise on more sympathetic planting schemes (Lucas, 1997) and in the 1970s a review resulted in a rise in the proportion of broadleaves planted (Aldhous, 1997).

In the latter half of the twentieth century our woodlands have seen other changes in emphasis. Increased awareness of the limited amount of woodland we have left, especially that which has never been cleared, together with better knowledge of the distribution and ecology of our native flora and fauna have made us aware of nature conservation and helped us value our woodlands for these reasons. The plight of tropical forests and the Rio Convention in 1992 reinforced this message and now biodiversity has become part of our language. There is also an increasing interest in continuing and restoring traditional management practices for historical and cultural reasons.

As the population becomes more affluent, the pressure on the land becomes greater and natural areas are increasingly valuable for recreation. As a consequence, many woodlands today are becoming increasingly multipurpose, integrating commercial forestry, nature conservation and recreation.

Recently another ambitious plan has been put forward; to double the area of woodland in England from 7.5 per cent in 1995 to 15 per cent by 2050. If successful, approximately 1 million hectares of new woodland will be planted returning the level of woodland cover to that last seen in 1086. These woodlands will be planted to reflect the current need for multipurpose forests.

WOODLANDS IN BRITAIN TODAY

Britain is one of the least wooded countries in Europe with only 10 per cent cover (Forestry Commission, 1996). Only Ireland and the Netherlands have less while France has 26 per cent and Sweden 68 per cent. However, the woodland that is found in Britain shows a remarkable variety. Figure 1.4 illustrates the broad routes by which an area of woodland may have originated and helps to define some of the terms describing woodland.

The *wildwood* was the natural woodland that developed in Britain as the ice sheet retreated and was largely unaffected by human activity. Most of the wildwood has been cleared. That which was not, has been managed and altered over the years. Two common forms of management were *coppicing* (Plate 1.2) and *pollarding* (Plate 1.3). In coppice, trees or shrubs are cut at ground level but shoot again to produce a crop of new

stems. Pollards are trees cut well above the ground, usually above head height, so the regrowing shoots are out of reach of grazing animals that are pastured on the land surrounding the trees. Both of these systems will be discussed in more detail in Chapter 3.

Land that has been cleared of its wildwood can become woodland again, either by planting with trees or by natural reversion to woodland. Open land that is not grazed will, in most areas, become colonised by trees. This process of woodland development (succession) is outlined more fully in Chapter 2, the result is usually called *secondary woodland* because the *primary* (or wildwood) has been cleared. Because these woods develop naturally, they usually consist of native species rather than planted introductions, and are referred to as recent *semi-natural* woods. Open land can also be planted with trees, consisting of either native or introduced species. This is usually done for commercial purposes and the woods are known as *plantations*.

The managed woods that were not cleared are known as *ancient woods* and, where they contain native species and do not have trees planted in them, they are known as *ancient semi-natural* woods. The management of these woods today is varied. They may be coppiced or pollarded, usually after a period of neglect. They may be managed as *high forest*, growing the trees on to their full height either from the old coppice stools or by cultivating natural regeneration. Many woods have not been managed in a specific way and have been neglected, but others have been purposely left and a policy of *minimum intervention* adopted. The term *natural woodlands* has also been used for this situation (Peterken, 1996).

Alternatively, ancient woods can be replanted with native or non-native species to create plantations (or *replanted ancient*

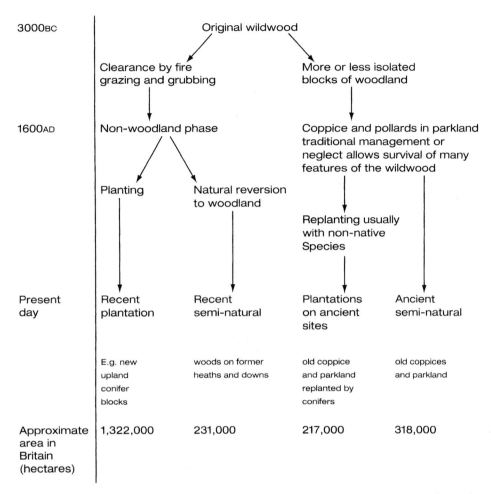

3000BC		Original wildwood		
	Clearance by fire grazing and grubbing		More or less isolated blocks of woodland	
1600AD	Non-woodland phase		Coppice and pollards in parkland traditional management or neglect allows survival of many features of the wildwood	
	Planting	Natural reversion to woodland		
			Replanting usually with non-native Species	
Present day	Recent plantation	Recent semi-natural	Plantations on ancient sites	Ancient semi-natural
	E.g. new upland conifer blocks	woods on former heaths and downs	old coppice and parkland replanted by conifers	old coppices and parkland
Approximate area in Britain (hectares)	1,322,000	231,000	217,000	318,000

Figure 1.4: The origins of British Woodland and their current areas. From Kirby *et al.* (1984), with figures for area from K. Kirby *pers. com.*

woodland). In most cases the existing trees are cleared before the new trees are planted but sometimes remnants of the ancient woodland can still be seen, for example when planting occurs round big old pollards which would have been too costly and time-consuming to remove (Barwick, 1996).

The system outlined in Figure 1.4 looks very clear and simple but complications occur. Although ancient woodland is usually defined as woodland that has occurred on the same site since 1600, there is always a possibility that an area was cleared of trees and replanted before 1600, as Rackham (1990) has shown. Not all ancient woodland is therefore derived from the wildwood.

The present-day management of woods may further complicate these distinctions. Recent planting is taking place which mimics the semi-natural woodland which is to be found (or predicted to be found) in the local area. While increasing use is made of local genetic varieties of tree species, some plantings have used native species of foreign origin.

Plate 1.2:
Coppice stools,
Essex

Plate 1.3: A pollard, Buckinghamshire

Further complications arise with the use of the word native. This term is usually applied to trees which colonised naturally after the ice age without human aid (see also Chapter 2). Native woodland in England and Wales usually refers to communities of native species that continue to regenerate but are not necessarily of natural origin. In Scotland, however, the term usually applies only to the remnants of original pine forest, which are naturally regenerating. In England and Wales this would be called ancient semi-natural.

Due to their importance (largely for different reasons), both ancient semi-natural woods and commercial plantations will be discussed more fully here. The other woodland types, processes and management are covered in the following chapters.

Ancient semi-natural woodland

Ancient semi-natural woods are the most important category of woodland for nature conservation (Peterken, 1993). The reasons why they are valuable are as follows (summarised from Peterken, 1996):

1 The tree and shrub communities may have descended from the wildwood.
2 The wildlife communities are usually richer than those of recent woods.
3 They may contain a high proportion of the populations of rare and vulnerable woodland species.
4 They may contain large old trees which may be several hundred years old and provide habitats for organisms characteristic of primeval woodland.
5 They may contain other natural features which are rare elsewhere, e.g. streams in their natural water courses.
6 They are reservoirs for other organisms e.g. woodland ride species, which may be rare in the surrounding land.
7 As they are less affected by disturbance they can be used as controls to look at the effects of people on e.g. soils.
8 They have been managed for centuries and provide a living demonstration of sustainability between humans and nature.
9 They may contain earthworks and ancient monuments worthy of preservation.
10 They are traditional features of the locality.

Point 2 above was well demonstrated by Kirby (1988a) who compared two adjacent woods less than 5 hectares in size in Essex. Both were formerly hazel coppices with oak standards. Plegdon Wood is pre-1600 in origin. Lady Wood was established in the nineteenth century on old agricultural land. Table 1.2 illustrates the differences in trees and ground flora.

In terms of the flora and fauna it is possible to identify specific plants and animals which, by their presence, usually show that a wood is ancient. These are known as ancient woodland indicators. Lists have been drawn up for a range of different groups (e.g. BTCV, 1997 and Marren, 1990 both for plants). The species used as indicators may vary according to which part of the country the wood is in. Some of the best-known ancient woodland indicators are plants such as the small-leaved lime (a species which was found extensively in the wildwood) and flowers such as bluebell and wood anemone.

It is not only plants that can act as ancient woodland indicators, some invertebrates do so as well. The black hairstreak butterfly whose larvae feed on blackthorn is one of the best (Peterken, 1993). There are examples from a wide range of organisms, including beetles, flies, molluscs and pseudoscorpions.

As well as the ancient woodland indicator species there are also indices of ecological continuity which can be calculated and used to give an indication of the degree of continuity of tree cover at or near a particular site. The index for lichens presented by Harding and Rose (1986) uses thirty species. It has the disadvantage that levels of atmospheric pollution also affect lichens. The saproxylic beetle index (Harding and Alexander, 1993) uses 195 species which enables it to be more sensitive, but the beetles are harder to find and may be active for only a short part of the season.

Apart from the plants and animals living in the woods, there are other ways, relating to their history, to distinguish ancient woods from recent ones. The boundaries of ancient woods are usually irregular and sinuous and may be associated with an earth bank and ditch. Ancient woods are usually marked on the first edition Ordnance Survey maps of 1800–1830 and may have names relating to old woodland terms or practice, such as coppice, wood or copse. Within these woods may be old coppice stools, pollards or old trees (see Marren, 1990 and English Nature, 1996 for more details).

Table 1.2: Comparison of the trees and field layer flora of two adjacent woods of differing history in Essex

	Plegdon Wood	Lady Wood
Age of woodland	Ancient	Recent
Type of wood	Coppice with standards pre-1600	Coppice with standards established nineteenth century
Size of wood	<5 hectares	<5 hectares
Basal area of trees in 12 plots in (m³):		
Total	103	125
Pedunculate oak	47	87
Ash	12	23
Birch species	3	15
Hornbeam	18	–
Wych elm	6	–
Field maple	12	–
Goat willow	5	–
Vascular plants in 10 × 10 m plots:		
Total species in 12 plots	51	33
Mean number of species per plot	17.3 ± 1.8	11.6 ± 1.0
Species markedly more frequent in one wood or the other (out of 12 sample points):		
Lesser celandine	11	1
Wood anemone	9	–
Meadowsweet	7	–
Germander speedwell	4	–
Male-fern	4	–
Lords and ladies	9	5
Wood-sedge	7	3
False Brome	1	8

Source: Kirby (1988a)

To facilitate the conservation of ancient woods, and using these types of criteria, the Nature Conservancy Council (now English Nature) have compiled an inventory of ancient woodlands over 2 hectares in size (English Nature, 1996 and Kirby *et al.*, 1984). From this, the amount of ancient woodland left in England, Scotland and Wales was established. Ancient woodland is still being felled, since the 1930s 7 per cent of that in England and Wales has been cleared and 38 per cent has been converted to plantation woodland.

The inventory established that about 340,000 hectares of ancient woodland exists in England. It might be expected that, because of their high nature conservation value, many woods would be protected by an official designation (such as SSSI or NNR) but 85 per cent have no such protection (Thomas *et al.*, 1997). Of all the sites only 14 contain more than 300 hectares of ancient woodland (or at least 200 hectares in a single block) and over 80 per cent are smaller than 20 hectares. The picture is thus of an extremely fragmented habitat with many woods surviving due to sympathetic management by their owners/ managers (or neglect) rather than legislation. The distribution across England is also patchy, with more than a quarter being found in Kent, Sussex, Surrey and Hampshire. Ancient woods can be very variable in their tree composition and form (e.g. coppice, pollard, or standard) depending upon location and historical management.

Modern Commercial Forestry

Although not usually of such high nature conservation interest, the areas of commercial forestry are important for economic reasons. In addition, they cover a large area and include a high proportion of the total woodland cover, see Figure 1.5.

The primary purpose of commercial forestry plantations is to make money by growing trees, felling them, selling the timber and then growing a new crop in their place (see

Chapter 3 for more details). There are many differences between commercial woodland and semi-natural woodland, plantations traditionally being large blocks of even-aged, single species crops. The terms hardwood and softwood are often applied to broadleaved and coniferous trees respectively in a forestry

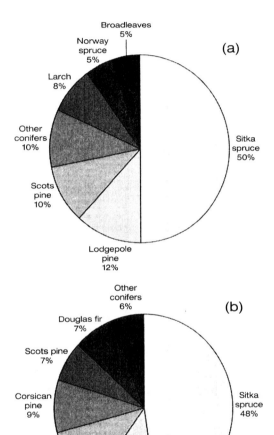

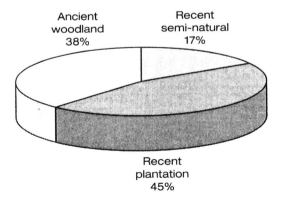

Figure 1.5: The amount of ancient woodland in England compared to that of recent origin (based on area). From Forestry Commission/Countryside Commission (1996).

Figure 1.6: The species composition of woodland managed by Forest Enterprise. (a) The tree species composition of existing plantations. (b) The tree species composition of currently planted plantations. From Forest Enterprise (1997).

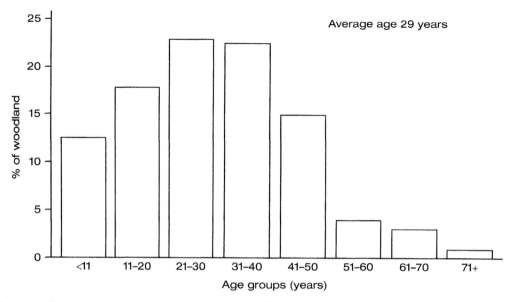

Figure 1.7: The age structure of trees on land managed by Forest Enterprise. From Forest Enterprise (1997).

Plate 1.4: A sawmill at Boat of Garten, Scotland

situation but this can cause confusion (Thurkettle, 1997) so the latter terms have been used throughout this volume.

Figure 1.6a shows the species composition of plantations planted by Forest Enterprise (now the commercial arm of the Forestry Commission) in the past and Figure 1.6b the species currently being planted. Some species have proved less successful than anticipated (e.g. Lodgepole pine), and others more productive (e.g. Sitka spruce). The amount of broadleaved trees has approximately doubled

Table 1.3: Uses of timber in Britain

(a) Main uses of hard and soft woods

	Production per year (m³)
Hardwood	
Sawmills	360 000
Paper and board mills	220 000
Panel mills	180 000
Firewood	150 000–900 000
Miscellaneous	240 000
Softwood	
Sawmills	2 000 000
Chipboard	
Orientated strand board (chipboard with stands aligned along length of board)	
Medium density fibre board	2 500 000
Cement-bonded particle board	
Paper and pulp	See Table 1.3b

(b) The British paper and pulp industry*

Mill	Product	Type of wood used	Quantity produced (tonnes per year)
Shotton Paper Co.	Newsprint	Spruce + 30% recycled fibre	440 000
Caledonian Paper Co.	Lightweight coated magazine paper		223 000
Iggesund Paperboard Ltd.	Packaging boards		172 000
St. Regis Paper Co. Ltd.	Fluting	Hardwood	125 000

Sources: Banks and Cooper (1997) and Thurkettle (1997)

Note: * These four mills use 1.6 million tonnes per year of small round wood and sawmill residues.

but still only comprises 11 per cent of the total planting. Most broadleaved trees grow more slowly than coniferous ones so the time taken to reach harvesting age is much longer, thus they are less productive. In the forestry industry as a whole (not just Forest Enterprise), broadleaves make up over 20 per cent of the commercial forest area, with oak representing nearly half of that (Forestry Commission, 1997). They are looked upon rather more favourably by the industry now than they were 70 years ago (Evans, 1997). The number of broadleaved trees planted annually in Britain (i.e. not just commercial forestry) now exceeds that of coniferous trees.

The age distribution of high forest trees usually shows very low numbers of older trees, most being harvested when the timber is of maximum commercial value. Figure 1.7 illustrates the ages of trees planted on Forest Enterprise land.

Harvested timber is converted in mills (Plate 1.4) and processing plants and then used in a wide range of industries, e.g. chip board, wood pulp for paper and, of course, as construction timber. Table 1.3 gives a break-down of current uses. The forest industry is now well established in Britain, although we still need to import some 90 per cent of our timber requirements. Awareness of associated issues such as nature conservation, landscape value and recreation are now more widely accepted by the foresters and increasingly efforts are being made in these directions. As Malcolm (1997) says, 'The next step is to develop appropriate silvicultural systems for ensuring a transition from simple plantation forestry to properly structured and ecologically sustainable forest.' The will to do this is now clearly established in the forestry industry.

2

THE ECOLOGY OF WOODLAND HABITATS

•

THE STRUCTURE AND FUNCTIONING OF TREES

Wood and trees

The growth of a tree, wood structure and function

A tree, according to Alan Mitchell (1974), is a woody plant able to exceed 6m in height and having a woody, usually single stem. The woody nature of the tree stems are due to lignin which stiffens the cell walls. Trees, like other plants, grow by the division of the cells in the bud meristems. This causes elongation and increases the height and length. They also have meristematic cells between the wood and the bark called the cambium. These cambial cells divide to produce xylem (or wood) on the inside and phloem (or bark) on the outside. Each year a new layer is produced between the old bark and wood. The outer layers of bark are usually shed so it remains a relatively thin layer, but the wood remains.

During the winter months trees in Britain hardly grow. In the spring they grow fast, producing springwood with large cells and thin walls. Later in the season they produce summerwood consisting of smaller cells with thick walls. These differences can be clearly seen as the annual rings when a tree is cut. The amount of wood added each year is dependent on the growing conditions and may also vary from one part of the tree to another.

Characteristics of mature and senescent trees

Trees naturally pass through several stages in their life. The young growth phase is when the tree is putting most of its energy into growing. As it reaches maturity some of the energy is shifted from growth to flower and seed production. The tree becomes over-mature when growth and seed production have declined and the amount of tissue where photosynthesis takes place (in the form of leaves) declines in relation to the amount of non-photosynthetic material (stem and branches). Different tree species show these characteristics at different points in their lives and some live much longer than others. It is not understood why there is so much variation in life span but the ageing process is very similar regardless of the species. As they get older, trees show a reduced growth rate, are slower to heal and the amount of die back in the canopy increases. This results in larger cavities in the branches and trunk, seepages, dead wood in the canopy, larger amounts of loose bark and a greater chance of physical damage. The over-mature stage in many trees can last a very long time; even centuries in species such as oak.

Decomposition of wood

The decomposition process may start when a branch falls off a tree or when a tree dies and remains standing, but it often begins when the tree is still alive. The dead wood in the centre of the tree provides structural support but if it starts to rot the process generally presents no problems to the tree itself. Decay commences when wounding causes a small number of cells to die, usually those in the inner bark. Agents such as wind or organisms like some fungi may enlarge the wound and the wood then starts to decay. It is also thought that some fungi may be present in the tree for many years but only become apparent when the tree dies or is stressed.

The decomposition becomes more extensive as fungi and wood boring insects break down the cell walls. Two broad types of rot occur, white rot when the lignin and cellulose are both broken down and red rot when the lignin is left intact. Invertebrates follow the fungi and reduce the heartwood to a soil-like material.

It is not just the upper parts of the tree that are subject to rot, there are various fungal species and invertebrates which hollow out or rot the roots too.

Evergreen and deciduous habits

In temperate regions, trees have two main strategies for growth. They are either evergreen and retain their leaves all year round or they are deciduous and shed their leaves during the winter months. The leaves are the food-producing zones of the tree. To assist with photosynthesis the leaves have pores (stomata) which allow the carbon dioxide (required by the tree) in and water vapour out. During the summer months, water is freely taken up by the roots, but in winter if the ground is frozen, this is not possible. In addition, in winter the daylight hours are short so the tree risks losing too much water from the leaves while not making enough carbohydrates to survive. To overcome this problem trees either have leaves which are very small (i.e. needle shaped) to reduce water loss, or lose their leaves during the winter. By losing leaves, deciduous trees may reduce their growing season and also reduce their levels of primary productivity. Table 2.1 compares various features for a conifer and a broadleaved tree growing within 1km of each other in Germany.

When deciduous trees lose their leaves in the autumn the leaves undergo the process of senescence or ageing. This tends to start at the ends of the leaves and progress towards their bases. When the green chlorophyll breaks down, the other pigments in the leaves, such as xanthophylls and carotenoids, become visible.

Table 2.1: Characteristics of a representative broadleaved and a coniferous tree

	Beech	Norway spruce
Age	100 years	89 years
Height	27 metres	25.6 metres
Leaf shape	Broad	Needle
Annual production of leaves	Higher	Lower
Photosynthetic capacity per unit dry weight of leaf	Higher	Lower
Length of growing season	176 days	260 days
Primary productivity (tonnes carbon ha-1 year −1)	8.6	14.9

Source: Begon *et al.* (1996)

Note: The trees were growing within 1 km of each other on the Stolling Plateau, Germany.

The result is the yellow and orange colours that are associated with the autumn months (Plate 1). As the leaves senesce, an abscission layer is formed between the leaf stalk and the branch. This is a thin plate of cells, all with soft cell walls, which form a weak zone eventually broken by the wind so that the leaf drops off. These processes occur in evergreen trees too but the leaves are not shed all at the same time like deciduous trees. Wareing and Phillips (1981) give more details.

Reproduction and survival of trees

Although all trees show the characteristics associated with the formation of wood, they come from a wide range of different plant families. Thus the form of their flowers and seeds can be very different from one type of tree to another.

Flowering and pollination

Most British trees flower in the spring. While those in more tropical regions may have large spectacular flowers, most of those found in Britain are smaller and often almost insignificant due to their method of pollination. Pollen, produced from male flowers (or male parts of the flower) usually needs to be transferred to another flower on a different tree (cross-pollination) where it will fertilise the ova and then produce seeds. In most species the male and female flowers are found on the same plant but in a few cases, such as holly and juniper, the bushes produce either pollen or the female berries.

Trees may use a variety of methods for pollination, the most common being by wind or insects. The former is most frequent in temperate regions, the latter in tropical areas. Thus, most British trees are wind-pollinated and produce large quantities of pollen early in the year, before the leaves of deciduous trees develop. One of the most obvious flowers of this type are the catkins on hazel which are visible soon after Christmas in most years. The flowers of wind-pollinated trees tend to be insignificant, without any smell, and are often high up on the trees to catch the wind. Most conifers are wind-pollinated as are oak and beech.

Some trees found in Britain are insect-pollinated, for example sycamore, others are largely insect-pollinated but show some characteristics of wind pollination as well (Packham et al., 1992) e.g. willows and limes. Lime trees, in particular, have a strong fragrance and are visited by large numbers of insects. Generally, these trees flower later in the spring and in early summer when there are more insects around. Figure 2.1 illustrates the flowering times of some tree species as determined by the presence of pollen 'caught' at a range of sites in England and Wales throughout the year.

Seed production and dispersal

British trees have seeds that are dispersed by the wind or animals. Wind-dispersed seed is usually light and may have special wings e.g. sycamore, ash and lime. Birch seeds have only slight flanges to catch the wind but the quantity of these light seeds is enormous and they spread very readily. Seeds reliant on animals for distribution are quite different. They are usually large and heavy and fall under the tree producing them or alternatively they are obvious berries remaining on the trees after the leaves have fallen. They all contain something of value to animals and are either eaten, or moved and stored for eating later (e.g. oak and beech) or the fruit eaten and the seeds dispersed in the droppings (e.g. rowan).

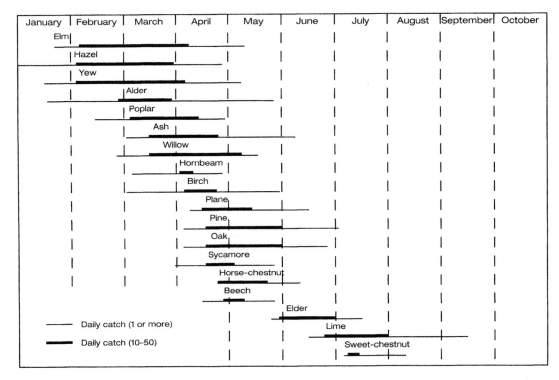

Figure 2.1: The flowering periods of British trees, determined by pollen grains in the atmosphere at a range of stations across England and Wales. From Godwin (1975).

Trees are perennials (they live and flower for many years) but they may not produce seed every year. One characteristic feature of some species is that they have mast years. The word mast originates from the Scandinavian mat meaning animal fodder because pigs were fed on acorns and beech nuts in the past (Evans, 1988). Mast has been defined as the periodic synchronous production of large seed crops (Nilsson and Wastljung, 1987) which means that in certain years the trees produce lots of seed but in the intervening years there is very little formed. Beech (Species Box 2.1) is perhaps the classic British species to which this applies. Between 1980 and 1995 the variation in mast production between years was found to be significantly greater than between different sites across Britain and five good mast

years were recorded (Hilton and Packham, 1997). It was not possible to predict good years (although at least one bad year always follows a good year), however, high temperatures and long sunshine hours the previous year (when the flower buds are forming) do increase the chances of a good crop (Matthews, 1955). Trees in clumps produced better crops than isolated trees, probably as beech needs to be cross-pollinated and this is less likely to happen if the tree is isolated. Table 2.2 shows the characteristics of seed production for some British tree species.

Some trees do not need to produce seed in order to regenerate. Several species are able to send out suckers from the main stem, e.g. elm and wild service tree. They are not able to colonise areas a long way from the parent tree but

Species Box 2.1: Beech

The beech tree is one of the most elegant of native trees. The classic vision of a beech is a slender, grey and smooth-stemmed trunk bearing spreading, pointed branches with oval, entire leaves. The beech is a shallow-rooted tree, found on well-drained soils and is native to the south of Britain. The leaves emerge in May, a fresh bright green and mature to a darker colour during the summer. In the autumn they turn a range of yellow, orange or copper colours before falling, making beech woods popular places for autumn walks. The over-wintering buds are narrow and pointed and a chestnut brown colour. The male flowers are in soft heads which are pale tan, the stamens being clearly visible. The female flowers are cup-like and stand up on the branches. The fruit is a three-sided nut, two of which are enclosed in a capsule. The quantity of viable nuts produced each year is variable and irregular, depending on complex climatic requirements. In most years the majority of nuts are empty. The nuts are edible and have been used in foodstuffs and cosmetics. The wood of the tree is fine-grained and strong and has been widely used in the

construction of furniture. It is less regularly used for outside purposes. Beech trees tend to prefer well-drained soil but can be found on both strongly calcareous soils and acidic ones.

Source: White (1995)

Table 2.2: Seed production of broadleaved trees in Britain

Species	Minimum seed-bearing age (years)	Interval between large seed crops (years)	Age when seed production declines	Time of seed fall
Alder	15–25	2–3	–	Sept–Mar
Ash	20–30	3–5	80	Sept–Mar
Beech	50–60	5–15	160	Sept–Nov
Birch	15	1–3	60	Aug–Jan
Hornbeam	10–30	2–4	–	Nov–Apr
Lime	20–30	2–3	–	Sept–Nov
Oak – Pedunculate	40–50	3–6	140	Nov
Oak – Sessile	40–50	2–5	140	Nov
Sycamore	25–30	1–3	70	Sep–Oct

Source: Evans (1988)

in the case of small-leaved lime it is an important ability, enabling the species to continue to survive in woods when little or no viable seed is set.

Germination and seedling survival

Although trees may produce huge quantities of seed, only a few will usually germinate. Worrell and Nixon (1991) recorded that in good seed years up to 700,000 acorns fell per hectare but approximately only 0.5 per cent survive to produce first year seedlings. Many fall prey to animals and the conditions also have to be right for successful germination to take place. If the seed is lying on the surface of the ground it runs the risk of drying out but being buried too deep is also detrimental. The fate of ash seeds produced in a Derbyshire wood is illustrated in Figure 2.2.

In natural systems the seed that germinates in a woodland mostly comes from the local trees. However, in commercial forestry, seed is sometimes collected for growing on and planting out, from selected trees. The reason for this is to ensure an adequate density of trees and also to ensure that the trees have the desired features for timber production. However, the characteristics selected (see for example Hart, 1991, Figure 4) may not be so beneficial from the nature conservation point of view. This is because they tend to be more uniform (so there is less genetic variation) and are less likely to show features such as abundant epicormic growth and those which encourage rotting to take place (which are beneficial to certain invertebrates and fungi – see section on invertebrates).

Good growth of seedlings often occurs after a good seed year, known as *brosse* in French oak woods (Evans, 1988), but in others almost none survive. Once the seed has germinated, it

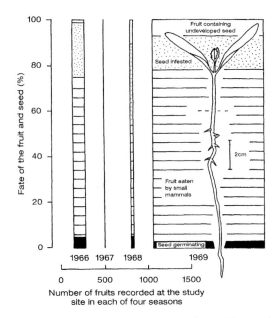

Figure 2.2: The fate of ash seeds in a Derbyshire wood during the period up to germination. The infested seed was damaged mostly by moth caterpillars. Numbers of fruits recorded are per m². From Packham *et al.* (1992).

has to compete with other plants growing on the woodland floor for light, water and nutrients. It also has to survive predation by mice and voles, which are partial to tender seedlings, and attack by fungi. Often, the main constraint for the young growing tree is the amount of light penetrating the tree canopy above it.

Shade-tolerant trees such as beech have the ability to 'sit', growing almost negligible amounts for many years, while the canopy above is dense. If one of the large trees nearby falls or dies, the saplings take advantage of this and can grow quickly up into the gap. Other species, like birch, are far less shade-tolerant and while they are often able to colonise gaps quickly (from seed), the saplings are not able to survive long in deep shade.

Mycorrhiza

Most, if not all, trees form mycorrhizal (from the Greek, fungus root) associations with fungi. The relationship between the fungus and the tree roots seems to be extremely close and symbiotic (both parties gain from it). There are two major types of mycorrhizal fungi. The ectomycorrhiza, or sheathing forms, cover the outside of the tree roots and may cause them to branch in a characteristic way. The fungus extends between the cells of the roots and obtains carbon as carbohydrates from the tree. From the fungus the tree receives minerals such as phosphorus, nitrogen and calcium, and also water. This type of mycorrhiza is very common in tree species. The roots infected with the mycorrhiza (Plate 2.1) can be found in the leaf litter around the trees but the fungus itself is more obvious when it produces fruiting bodies. Some of our best known toadstools are mychorrizal, examples being the fly agaric (Species Box 2.2) which occurs on birch and the amethyst deceiver on beech.

Vesicular arbuscular (VA) mycorrhizae live in the roots of the tree, between the cells of the host and penetrating into them. They do not alter the visible structure of the roots and have microscopic fruiting bodies so are difficult to see. Again the tree provides carbon for the fungus which transfers phosphate in return. When phosphate is limited, the presence of mycorrhizal fungi increases the growth rate of the plant but when phosphate is readily available the fungus may actually decrease the growth. This form of mycorrhizal association is found in ash and maples. Some trees (e.g. alder, willow and poplar) may have both types of mycorrhizal fungi, even in the same root (Hart, 1991). Mycorrhizae seem to be most abundant when mineral deficiencies would otherwise limit the growth of trees and are least abundant in very fertile soil and under dense shade (Kozlowski *et al.*, 1991). In some areas it has proved impossible to grow trees without their mycorrhizal-forming fungi. Innoculation of the fungus may be necessary when trees are grown in nurseries, especially in sterilised soil, and in areas of

Plate 2.1: Beech roots (laid on beech leaves to enable them to be seen more clearly). The left hand example shows roots that have an ectomycorrhizal association, the right hand example does not.

Species Box 2.2: Fly agaric

The fly agaric with its white spotted red cap is the classic fairy toadstool. The name 'fly' is given because the cap used to be broken into milk to kill flies. It stands up to 180 mm high on a white stem and has a volva, or white membranous material attached to the base, and also a ring towards the top of the stem. The white spots are actually lumpy warts, which are washed off in the rain (along with some of the red colour). The toadstool is often eaten by slugs and snails. Under the cap are white gills producing white spores. The fly agaric is always associated with birch trees with which it forms a mycorrhizal association. It is poisonous to people.

Source: Phillips (1981)

---

industrial wasteland which are being renovated.

Complex interactions may occur between different species of plant. Problems have occurred during afforestation in Britain when sitka spruce has been planted on heather-dominated (*Calluna vulgaris*) sites. The spruce shows a check in growth and symptoms of nitrogen deficiency. It seems that the spruce are unable to obtain nitrogen because their mycorrizal activity is suppressed by something produced by the heather roots (Robinson, 1972).

Allelopathy

Some plants appear to produce compounds which inhibit the growth of others. These effects have been termed allelopathic. Such compounds can be produced by any part of the plant but higher concentrations are found in the leaves and fruit. Some tree spe-cies such as black walnut have been shown to produce exudates which suppress the growth of other plants underneath them despite having a fairly sparce canopy (Packham, *et al.*, 1992). While many inhibitory effects can be demonstrated with different species it is not clear how important these are in natural ecosystems (Kozlowski, *et al.*, 1991).

Nitrogen fixation

A few species of tree (e.g. alder) have nodules on their roots containing bacteria-like organisms that are able to extract nitrogen from the air. This enables the trees to grow in nitrogen-deficient areas and they are often used in rec-lamation work because of this. Leguminous plants are perhaps the best-known nitrogen fixers and this is true of leguminous trees too. Exotic species such as *Robinia* are frequently used for this purpose.

THE STRUCTURE OF WOODLAND

The physical structure

General

Woodland is very much a three-dimensional habitat. The sheer height of the trees means that there is scope for many layers of plants underneath. While some types of woodland (e.g. beech) may have a very sparse covering of ground flora, others, especially along the woodland edges, may have a very complex structure.

Conventionally the plants are divided into four main layers. The canopy is the tallest layer and, while sometimes there is variation in the height of individual trees, they can also be more or less equal. The shrub layer or understorey is the woody component under the trees and is composed of shrub species and tree saplings. Beneath the shrub layer is the herb or field layer, consisting of plants that may die down in the winter or, like bramble, persist all year round. On the ground layer are the mosses and liverworts. These layers may not be particularly distinct and one may blend into another (see Figure 2.3). The age structure of the trees themselves often adds to the diversity of a wood.

It is important not to forget the underground parts of a wood. This is where the roots of the plants extend. Some tree roots

>5m
Canopy

2–5m
Shrub

0.1–2m
Field

<0.1m
Ground

Figure 2.3: The structure of a typical wood.

extend to deep levels but others are surprisingly shallow and spread out into the leaf litter layer forming an extensive network, extended even further by the mycorrhizal fungi.

Where the canopy is continuous, the ground underneath often consists mainly of leaves in various stages of decomposition. In places where sunlight reaches, the shrub layer is likely to be more extensive. Thus, it is largely the tree density which governs the horizontal variation in the woodland structure.

Larger gaps in the original forests may have been maintained as open spaces for long periods of time by large herbivores, particularly deer. In these open areas herbs with flowers requiring insects for pollination are found and these are the communities now associated with woodland rides.

The structure of the wildwood

There are probably no woods remaining in Britain that have been left untouched, the nearest being scrubby fragments on cliffs and in gorges (Kirby, 1988a). So to get an impression of what the wildwood might have looked at we have to look at other countries. Peterken (1996) gives a map of the virgin forests (i.e. those never significantly influenced by people) in Europe and discusses them in detail. One of those described is Białowieża Forest, in Poland, which is dominated by hornbeam, lime and oak with swampy alder and ash forest and tracts of conifers. Despite being sometimes claimed as an untouched wilderness, Peterken shows how even this area has been 'substantially affected by human activity'. It has been used in the past for hunting and the subsequent fluctuations in the number of large herbivores have affected tree regeneration. Nevertheless, Białowieża can perhaps give us a better idea of what one type of true virgin forest might look like and Fuller (1995) summarised the seven main features he observed there:

1 The trees are very high, commonly more than 35m, but the trunks are usually less than 1m in diameter.
2 There are huge quantities of dead wood, both standing and fallen.
3 There is little field or shrub layer except in wet areas or where a tree has fallen making a gap allowing dense regeneration (but good soil quality may encourage the growth of a field layer, K. Kirby pers. com.).
4 The trees form several layers of canopy.
5 There are a large number of tree fall gaps, ranging from single tree sized to 200 m x 800 m, all created by storms, but no permanently open areas.
6 Some patches of forest are very uniform in age structure, probably due to natural regeneration in a tree fall gap.
7 Extensive areas are swampy or periodically flooded.

Variation in structure over time

After the last ice age, trees progressively colonised Britain in a process known as succession. Pioneer shrubs came first and then tree species good at colonising open ground, e.g. birch. These types of trees have abundant small seeds and can grow quickly. Late successional trees then follow on. These are species like oak and beech that have larger seeds and grow more slowly, but eventually, will shade out the earlier trees. The succession of vegetation types growing following the retreat of an ice sheet has been studied in depth at Glacier Bay in Alaska and is summarised by Begon et al. (1996).

In the early part of the twentieth century, ecologists (notably Clements) believed that succession led to a climax vegetation, the end product being characteristic of the climate of the particular area, and that this climatic climax was relatively stable. It has been increasingly realised that the situation is more complex. For example, climax is not solely determined by the climate, factors such as soil type and moisture levels are very important too. Also, the climax is not as stable as first thought, both small-scale and large-scale disturbances occur, making it a much more dynamic system.

When an existing area of woodland is cleared of its trees or an area of open land which has been treeless for many years is left, trees usually colonise in the same order, early successional trees and then late successional ones. This secondary succession can occur over a large area, for example, in Britain when a heathland is left ungrazed or uncut. It can also take place on a much smaller scale within woodland when some sort of disturbance causes a gap in the canopy. If a single tree falls, the gap may not be very large and the remain-ing trees will just extend their branches to make use of the extra light. A bigger gap may allow a succession to take place. This can be seen in Plate 2.2 where a group of silver birch is growing in a beech wood in a gap created several years previously.

Adaptations to woodland life

Vascular plants

The growing conditions for plants under the tree canopy are very different from those out in the open. Light levels are severely reduced, either all year round under conifers or during the summer months under deciduous trees. While this is a major problem for plants that need light for photosynthesis, there are some definite advantages to living under trees. Shaded areas tend to be moister and, close to the woodland floor, carbon dioxide levels are increased due to the decomposition processes taking place there. Both water and carbon dioxide are essential for plant growth. Under

Plate 2.2:
Regeneration of birch trees in a canopy gap

the trees it is also sheltered and where there is a gap in the canopy this can produce patches which can be hot and sunny. Patches of sunlight penetrating the canopy for short periods of time are called sunflecks; large sunflecks where the sun passes over a semi-permanent gap are timeflecks. The dappled effect in woods which ripples and changes as the wind blows the leaves is made up of windflecks which alternate between sun and shade at a rate of about 20 cycles per second. Plants growing in these conditions need to adapt to make the best use of the situation and to survive. Four kinds of species are considered here:

1 shade-tolerant species;
2 vernal (Spring) species;
3 species avoiding the need for light;
4 non-adapted species.

Shade tolerant

These plants have modified their leaves by increasing the size of them and the amount of chlorophyll they contain. They also have a thinner cuticle and larger gaps between spongy mesophyll cells in the centre of the leaf. They tend to have fewer layers of leaves and these are held more horizontally than other plants. Good examples of this are enchanter's-nightshade and herb-paris. A few plants like yellow archangel can grow in the shade or in more open situations, such as hedgerows, and have different types (and shapes) of leaves (shade leaves and sun leaves) according to where they are growing.

Vernal species

In deciduous woods there is a short period when conditions are warm and moist, and therefore suitable for growth, when the trees do not yet have their leaves. Some species take advantage of this short window and grow quickly. Bluebell (Plate 2 and Species Box 2.3) is the classic example of this type of strategy. In the deciduous woodlands of mainland Europe a progression of different coloured flowers appear in spring. These communities which occur in the same place at different times are called aspect societies (see Figure 2.4). Aspect societies occur in the British flora but are not usually so extensive.

Species avoiding the need for light

A few plants have become so completely adapted to shade that they do not need to photosynthesise and hence lose their green colour. This can be achieved in one of two ways. Either they become saprophytic, feeding on the decaying leaves of the woodland floor, an example being the yellow bird's nest. Alternatively, like toothwort (Plate 2.3), they may become parasitic and feed on the roots of the trees.

Non-adapted species

Finally, some 'woodland' species are not adapted to shade conditions at all but survive by growing in clearings, rides and round the edges of woods. Many of these were once much more widespread in the countryside but are now largely associated with woods.

Strong adaptations to woodland conditions can present problems to plants if the conditions change. Rackham (1975) observed sun scorch in dog's mercury (Species Box 2.4) when a large sunfleck after rain let more light down to the ground than the plants were used to.

The plant species found in ancient and well-established woods may show quite different characteristics from those that inhabit recent

Species Box 2.3: Bluebell

The bluebell is perhaps the most typical British woodland spring flower. It forms dense carpets of 4–16 blue, bell-shaped flowers on a single stalk between April and June. The leaves start to grow during the winter months from a deeply buried bulb. They reach the surface of the soil in January or February but do not grow further until the weather warms up in March. After flowering, the seeds are produced and, while the seed heads may persist into August, the leaves are usually dead by July. The leaves may grow to about 0.25 m, the flowers are taller at 0.45 m. The main method of regeneration is by seed. These may germinate from October onwards with some surviving into the spring before germination takes place. It takes about five years for a young bluebell plant to start flowering. Bluebells are mostly found in broadleaved woodland, but they do occur in conifer plantations, scrub and occasionally on heathland and moorland. Their basic strategy is to grow and reproduce before the leaves come on the trees and cast a dense shade over the ground. The number of leaves produced by each bulb is predetermined during the autumn and winter period, thus the plants cannot produce more leaves if they become damaged and are very susceptible to trampling and grazing.

Source: Grime *et al.* (1988)

woodland, see Table 2.3. Woodland plants may also be slow in colonising new areas. It has been found that it may take around 250 years for species such as bluebell, dog's mercury, primrose and sanicle to appear in new woods (Rodwell and Patterson 1994). Interestingly they may also persist for a surprisingly long time after the wood is cut down and can also occur under other plants (e.g. bracken) that cast a dense shade during the summer months, for example on Blackdown in the Mendips.

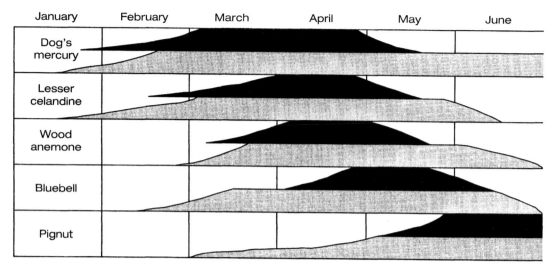

	January	February	March	April	May	June
Dog's mercury						
Lesser celandine						
Wood anemone						
Bluebell						
Pignut						

Figure 2.4: Aspect societies. The periods of flowering and vegetative growth of dominant ground flora species in a Hertfordshire wood. Black represents the flowering period and the stippled area the period of active vegetative growth. From Tansley (1939).

Plate 2.3: Toothwort, Wiltshire

Mosses, liverworts and lichens

These 'lower' plants live in two quite different habitats. The first is on the woodland floor, old tree stumps, rocks, etc. and the second is as epiphytes on the trees themselves. Due to their requirements for high humidity, most mosses and liverworts live in the shady areas of woodland, especially where it is damp, e.g. close to streams. They are most abundant in areas of high rainfall. Opening up of the canopy lets in too much light and heat and many species are unable to re-colonise easily, thus the best woods are those which have never been coppiced or felled (Marren, 1990). However, some species such as the distinctive cushion moss (Species Box 2.5) are able to thrive in dry woodlands.

Epiphytic lichens (Plate 2.4) have rather different requirements and tend to live in more open places receiving plenty of light. Good places are along woodland edges, high in the canopy or on more isolated trees. Some of the

Species Box 2.4: Dog's mercury

Dog's mercury is a shade-tolerant herb found on woodland floors, growing to 0.4 m. It is found on neutral to calcareous soils throughout the British Isles except northern Scotland but is rare in Ireland. In upland areas it may be found in less shaded situations. The green flowers are produced from February until May. The green hairy and oval leaves are fully expanded by May but may persist until late winter. The seeds set from June to August in a two-seeded capsule and are shed explosively. They germinate the following spring but most of the regeneration is probably by vegetative means. The plants are either male or female and have extensive underground rhizomes and form large patches consisting of the same individual.

Source: Grime *et al.* (1988)

best woods for lichens are wood pastures where old trees are well spaced. Many lichens are dependent upon certain tree species and some also need old trees and a continuity of mature wood to be able to survive, see Table 2.4. Britain seems to have a better epiphytic lichen flora than much of the rest of northern Europe (Kirby, 1988b) despite the sensitivity of lichens to pollution. Perhaps this is because of the relatively high number of old trees we have and our prevailing westerly winds which usually bring clean air (K. Kirby, pers. com.). Even some of the most sensitive species to pollution (*Lobaria spp.*, Species Box 2.6) can be found in the humid west. Interestingly, lichen-rich sites may not necessarily be rich sites for higher plants but they are often good for other groups of organisms (Gilbert, 1977, cited in Kirby, 1988b).

Fungi

Fungi play a fundamental role in woodlands. Through their role as mycorrhizal species they help trees to grow and they also have a major part to play in the breakdown of dead organic

Table 2.3: Characteristics of plants typical of old and young woodland

Old woodland species (e.g. wood anemone, herb-paris)	*Recent woodland species (e.g. rosebay willowherb, bramble)*
Need a stable environment	Are opportunists and quick to colonise disturbed areas
Are confined to woods	Also found in hedgebanks, field corners and wasteground
Are poor colonisers	Are good colonisers
Produce few seeds	Produce many seeds
Have large seeds which fall close to the parent plant	Have small light seeds, often transported by wind, or berries
Are slow-growing	Are fast-growing
Are long-lived	Are often biennials or annuals
Have no persistent seed bank	Can lie dormant as buried seed
Are adapted to shade	Are indifferent to shade
Often grow in localised patches	Are common and widespread
Are often poisonous	
Thrive under traditional coppicing regimes	Thrive under disturbance and modern forestry methods
Are declining	Are increasing

Source: Marren (1990)

matter; both leaves and wood. The fungal mycelium (the growing and feeding part of the fungus) is present all year round, and can be very extensive, but it tends to be the spore-producing fruiting body that is more noticeable. Many of these we associate with the autumn months when it is damp and still relatively warm. Fungal fruiting bodies found on the woodland floor come in a variety of shapes including the classic toadstools and earth balls. On trees and standing or fallen dead wood the bracket fungi are very obvious e.g. birch polypore (Species Box 2.7) and some of these are perennial and last for many years.

Some of the smaller fungi are especially beautiful like the black and white candle snuff and the slime moulds (a different group to the true fungi, Species Box 2.8) which can be like miniature lollipops in shape. Different types of woodlands have characteristic fungal flora e.g. upland birch woods and native Scottish pine woods have specific species associated with them, which are now quite rare.

Invertebrates

It is extremely difficult to summarise the adaptations of invertebrates to the woodland habitat in one short section. Woodlands are the richest invertebrate habitat with as many species in Britain (15,000) as in all other habitats put together (Marren, 1990). Amongst woodland specialists there are those which are dependent on the structure of the woods and also many which have specific requirements for a particular species of tree or even a microhabitat on that type of tree.

Invertebrates have two characteristics not commonly found in other groups of organisms. They are often annual, so detrimental conditions can quickly lead to extinction of the entire population. Second, the larval stages and the adults often live in different places and can have very different requirements.

Herbivorous insects may have very definite food preferences, for example the caterpillars of butterflies and moths need a good colony of their food plant to be able to support a viable population. Table 2.5 shows the food plants

---

Species Box 2.5: Cushion moss

Leucobryum glaucum is one of the few mosses that can be easily recognised, even at a distance. The plants themselves range from less than 30 mm to more than 100 mm but they grow in tight cushions which are a pale, grey green colour and are very conspicuous. They often form extensive areas of moss on the woodland floor offering a contrast to the brown leaf litter. The cushions easily become detached from the woodland floor and if turned over often continue growing, so forming complete balls. Cushion moss is only found in acidic areas growing under oak, beech or conifers but it also grows on moorland. Although small capsules producing spores are formed, they are rare and the most usual form of reproduction is vegetative. Purple-coloured rhizoids arise in the centre of a group of leaves from which new plants appear.

Source: Watson (1981)

---

required by different woodland species. Many woodland butterflies like sunny areas to fly in and prefer woods with plenty of rides and clearings.

Other plant-eating insects include the beetles and true bugs. Many of these feed on the trees themselves, with some trees supporting a wider range of species than others (see Table 2.4). As well as the plant-eating species there are invertebrate predators that specialise in hunting for food on trees. Some are widespread in the canopy, including the tiny spider *Theridion pallens* (Species Box 2.9) and the oak bush cricket that, unusually for a cricket, eats other insects (Species Box 2.10). Other invertebrate predators or parasites are specialists living in very precise places, like the spider *Drapetisca socialis* which lives at the bases of beech trees or some of the rot hole breeding flies which need exactly the right kind of rot.

The figures in Table 2.4 do not include the saproxylic insects, which are associated with dead wood. The larvae of a range of flies and beetles are dependent on rotting wood and the fungi breaking it down. Some of these are specialists on particular tree species and even the

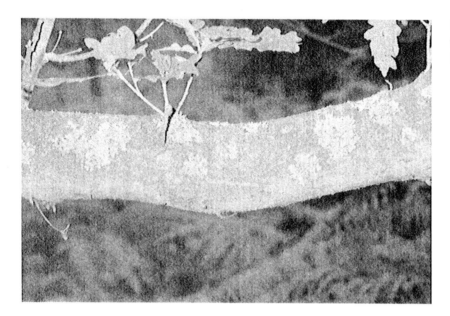

Plate 2.4: The lichen community on an oak branch

Table 2.4: The number of lichens and plant-eating insects associated with different tree species

Tree or shrub	Number of lichens	Number of plant-eating insects†
Oak (pedunculate and sessile)	324	423
Ash	255	68
Beech	206	98
Elm species	187	124
Hazel	160	106
Willow species	160	450
Scots pine	132*	172
Birch (downy and silver)	126**	334
Rowan	125**	58
Alder	105	141
Holly	96	10
Field maple	93	51
Lime species	83	57
Hornbeam	44	51

Sources: BTCV (1997) and Kennedy and Southwood (1984)

Notes:
†   Not including saproxylic species
*   In Caledonian pine woods
** Mainly in upland woods

Species Box 2.6: *Lobaria* species

Lichens consist of two different organisms in a symbiotic relationship. The first is an alga, which is able to photosynthesise and the second is a fungus. The resultant structures have no true roots and no protective cuticle so they grow when conditions are wet, but dry up when the weather is dry and sunny. Hence they grow best in the wetter west of Britain. Many different lichen species grow on trees and some are quite difficult to identify. They range from tiny encrusting forms, hardly recognisable as lichens, on the trunks of the trees to extensive mats on horizontal branches. One of the most spectacular of British genera is *Lobaria*. There are five species found in Britain, all of which are woodland species and are now largely confined to the west and northwest. The air is less polluted here and *Lobaria* is very sensitive to pollution. It is a relatively large thallose lichen with lobes up to 0.25 m long. Despite the generally cleaner air in Britain now, *Lobaria* still has a relatively restricted distribution because it has poor recolonisation abilities. *Lobaria pulmonaria* is also called lung wort

and in the past it was sold as a cure for lung diseases.

Sources: Dobson (1992), Richardson (1981)

Species Box 2.7: Birch polypore

One of the commonest bracket fungi is the birch polypore, sometimes called the razor-strop fungus. The white brackets form on dead birch trees and start out globular in shape, develop into a hoof shape and then enlarge to a classic bracket 100–200 mm across. As they get older they also become greyish brown in colour rather than white. The spores are produced from pores on the underneath of the bracket. The fruiting bodies can be found all year round, they are annual but may remain on the trunks of the trees for longer. Often several fruiting bodies are found on the same tree.

Source: Phillips (1981)

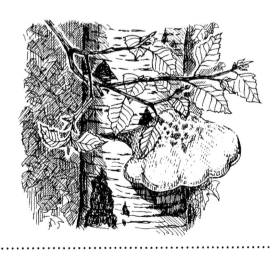

··

Species Box 2.8: Slime moulds

Slime moulds are extremely common organisms in woodlands, but because of their small size, are rarely seen. For much of its life, a slime mould consists of a gelatinous mass, which moves around slowly in or on dead wood and decaying matter, consuming microscopic food. At this stage it is called a plasmodium, and is quite animal-like in its characteristics. The plasmodium then transforms into the spore-bearing phase or sporophore, which is much more plant- or fungi-like in character. The sporophore is the more obvious part of the life cycle (though often still tiny in size) and is quite variable depending on the species. Some sporophores are large, amorphous-looking 'blobs' with a protective covering over a mass of spores. These can be bright and colourful, ranging from white and crumbly-looking, to a smooth silver colour or bright yellow. Other sporophores are tiny, with several structures being found close together.

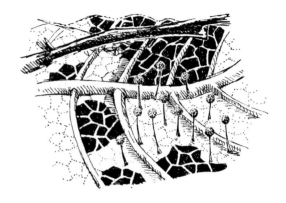

These are often lollipop-shaped, a near perfect round ball on top of a thin stalk, or 'balls' sat straight on a log. They occur in a variety of shapes and colours.
Sources: Feest (1996); Katsaros (1989); Stephenson and Stempen (1994)

··

precise type and stage of rot. The stag beetle (Species Box 2.11) is perhaps the largest and most spectacular of our saproxylic fauna. Another host of invertebrates is at work on the woodland floor breaking down the leaves that have fallen there and these will be discussed in a later section.

In summary, a good wood for invertebrates probably has a mixture of native tree species and a variety of tree ages present, including some over-mature trees with plenty of dead and decaying wood on them. It would have lots of large pieces of dead wood on the ground and good structural variety with clearings and rides, and a range of flowering herbs and shrubs.

Birds

Woodlands are extremely important habitats for birds, supporting more species of breeding bird than any other habitat in Britain (Fuller, 1982). The densities of birds breeding can also be higher than in any other terrestrial habitat (Fuller, 1982). However, very few rare birds are woodland dependent (only 14 per cent of those species with national populations of less than 10,000 pairs (Fuller, 1995)). Broadly speaking, there is rarely a close link between the bird species found and the tree species (although those commonly associated with broadleaved or coniferous woods can be distinguished), it is the habitat structure which is more important. Unlike other groups of organisms the recent management of a wood tends to be more important than the past management. Some deciduous woodland birds are listed in Table 2.6 together with their habitat preferences.

In addition to a good food supply, birds also need plenty of suitable nesting sites. Some woodland birds such as the woodpeckers

Table 2.5: Woodland butterflies and their larval food plants

Species	Larval food plant	Flight time
Dappled shade		
Green-veined white	Cresses	Apr–Jun, Aug
Silver washed fritillary	Violets	Jun–Aug
Speckled wood	Grasses	Apr–May, Jul–Sept
White admiral	Honeysuckle	Jun–Aug
Newly cut woodland and rides		
Heath fritillary	Cow wheat/plantains	Jun–Jul
High brown fritillary	Violets	Jun–Aug
Pearl bordered fritillary	Violets	May–Jun
Canopy species – tree and shrub feeders		
Black hairstreak	Blackthorn	Jun–Jul
Brown hairstreak	Blackthorn	Aug–Oct
Purple emperor	Sallow	Jul–Aug
Purple hairstreak	Oak	Jul–Aug
White-letter hairstreak	Elm	Jul–Aug
Woodland margins		
Brimstone	Blackthorn/alder buckthorn	Mar–Jun, Aug–Oct
Comma	Nettle/hop/currant	Mar–Jun, Jul–Oct
Duke of Burgundy fritillary	Cowslip	May–Jun
Gatekeeper	Grasses	Jul–Sep
Holly blue	Holly/ivy	Apr–May, Jul–Sep
Ringlet	Grasses	Jun–Aug

Sources: Lane & Tait (1990) and Marren (1990)

(Species Box 2.12), pied flycatcher and tawny owl (species box 2.13) nest in holes in trees so a good range of dead, decaying and hollow branches on trees are important and these are often lacking, especially in commercial forests. Other birds, nesting close to the ground, need good cover in the field and shrub layers which may be lacking if the wood is heavily grazed. Fuller (1995) provides a comprehensive review of birds in woodlands.

Woods have a very three-dimensional structure and different species of bird are found at different heights which can be linked to their food preferences and general habits, see Table 2.7.

Mammals

Many mammals use or live in woodlands but very few are entirely dependent on them. At the time of the wildwood the mammal population included several species of large herbivore which would have helped to keep clearings open; the beaver would have had pronounced effects on water courses and woods close by and wild boar would have had significant

Species Box 2.9: *Theridion pallens*

Theridion pallens is a tiny (1.25–1.75 mm), almost spherical spider, which is very common in the canopies of both deciduous and ever-green trees. While its size suggests that it might be a money spider, it is in a different family, that of the comb-footed spiders. It can be found at almost any time of the year, usually by beating the branches of the trees. It is variable in colour and markings but often looks quite pale and usually has white patches on either side of the front of the abdomen. The females lay their eggs in a distinctive white spiked sac, which they usually hide under a leaf and guard.

Sources: Locket and Millidge (1951); Roberts (1995)

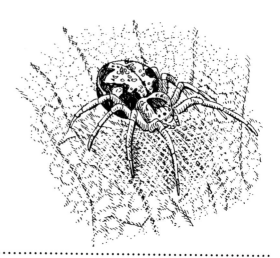

Species Box 2.10: Oak bush cricket

The oak bush cricket is almost entirely bright green in colour and is between 12–15mm long. The wings reach the end of the abdomen and the antennae are long, about four times the body length. The males have two pincer-shaped cerci at the end of the body, the females have a large, long (9 mm) ovipositor. The eggs are laid on tree trunks and hatch in June. They pass through five nymphal (larval) stages before becoming adult in late July or August. The adults may survive until the end of November. Oak bush crickets are arboreal and live only in the canopies of trees. While they prefer oak woods they are found on a variety of broadleaved species. They are noc-turnal and the only British bush cricket which is predatory. During the day they rest on the underside of leaves but at night they hunt for small insects such as caterpillars and aphids. They are attracted to light and are often found indoors or in moth traps. This cricket makes a purring noise by raising one of its hind legs and drumming the tarsus (end segments of the leg) on the ground or, more usually a leaf. Oak bush crickets can be very common in wooded areas

in the south of Britain but are scarce north of a line between the Dee and the Humber and also in Ireland.

Sources: Bellmann (1985); Marshall and Haes (1988)

Species Box 2.11: Stag beetle

The stag beetle is Britain's largest beetle at up to about 50 mm long. Both sexes are shiny black with deep red elytra (wing cases) but only the males have large antler-like jaws, which are used for wrestling with each other. The larvae are large white grubs living in the rotting stumps and roots of deciduous trees. They eat the wood and take three and a half years to reach the pupation stage. The adults emerge in the summer months and fly on warm evenings. Because of their size, the fact that they are attracted by light and have a rather clumsy flight they are very noticeable. Stag beetles used to occur over much of England and Wales but their range is now confined to the Thames valley and parts of Essex and Hampshire. They require dead tree stumps but these do not necessarily need to be in woodland. In fact a stronghold for this species is now the suburban area of south London. Despite now not being a strictly woodland species the

stag beetle is an important 'flagship' species for those requiring dead and decaying wood. Most other saproxylic beetles are much smaller, less imposing and less likely to catch the imagination of members of the public.

Source: English Nature (1994)

rooting effects. Today we have no beavers and domestic pigs forage in a very small handful of sites. We do still have large herbivores (though not necessarily the same species) in the form of deer. Table 2.8 gives notes on the six species of deer currently found in Britain. Most are widespread woodland species (e.g. roe deer, Species Box 2.14) and the impact of deer today can be quite problematic. They eat the new shoots as they start to grow from coppice stumps and are also a problem in commercial plantations where they browse young trees. While the presence of some deer is probably beneficial to the structure of the wood, in many instances the numbers are too high and the deer have a negative impact on commercial crops or on the natural regeneration in woods. Table 2.9 lists the main differences between woods with high and low numbers of deer.

Perhaps the most commonly seen mammal associated with trees is the grey squirrel, which is well adapted to an arboreal life style and to feeding on tree seeds. Grey squirrels were introduced into Britain from North America at the turn of the century. They occur in deciduous woodland throughout most of the country, where the indigenous red squirrel is no longer found. Grey squirrels have had a very significant impact on British woods in the time that they have become established. One of the main problems caused by squirrels is bark stripping. This can take place on a wide range of species but beech and sycamore are preferred. Bark stripping usually takes place between May and July on trees with vigorous growth and diameters between 10–25cm. The squirrels will also strip the bark off branches of larger trees. The damage wounds the tree and, if severe, will kill it. Mountford (1997) illustrated that squirrel damage can alter the long-term development of an unmanaged wood by killing large numbers of beech trees.

Table 2.6: Deciduous woodland birds and their habitat preferences

Species	Clear fell	Pre-thicket	Thicket	High forest	Over-mature	Comments
Coal tit				**	***	Prefers conifers. Needs cavities on or near ground for nesting.
Lesser spotted woodpecker				**	***	Feeds in canopy. Needs dead branches for drumming and excavating nest sites.
Pied flycatcher			*	**	***	Prefers mature and over-mature upland woods. Needs dead branches for perching and natural holes for nesting.
Wood warbler			*	**	***	Needs sparse understorey and good canopy.
Green woodpecker			*	**	***	Needs open areas to feed on ants and other invertebrates. Can excavate holes in live trees.
Great spotted woodpecker			*	**	***	Needs dead branches for drumming. Can excavate holes in live trees.
Marsh tit			*	**	***	Needs natural cavities for nesting. Most nests are below 1 m high.
Nuthatch			*	**	***	Needs cavities for nesting. Prefers mature and over-mature woods. Poor coloniser.
Redstart			*	**	***	Needs mature and dying timber for feeding and nesting. Sparse understorey for display flights.
Tawny owl			*	**	***	Wide range of woodland types. Needs dead branches and holes for nesting.
Jay			**	**	***	Prefers mature woodland with good understorey. Feeds in glades.
Treecreeper		*	**	**	***	Prefers mature, over-mature and dead trees.
Sparrowhawk			**	***	**	Prefers large woods or wooded landscape. Edge habitats are important.
Woodcock	*	**	**	***	**	Needs wet, invertebrate-rich soils for feeding. Dry areas for nesting.
Hawfinch			**	***	**	Uses the canopy but feeds on the ground in winter. Needs early spring buds and seeds.
Spotted flycatcher			*	***	**	Prefers woodland edges and glades.
Chiffchaff			*	***	**	Pefers mature woodland. Tall trees as song posts. Good understorey for nesting.
Blackcap		*	**	***	**	Prefers mature woodland with good canopy and understorey.
Turtle dove			***	**	**	Prefers thickets for nesting and open glades for feeding. Woodland edge plants are good food.

Wren	**	**	***	**	**	Wide range of woodland types. Prefers dense understorey.
Garden warbler		**	***	**		Prefers open woodland with good field and understorey for nesting.
Long-tailed tit		**	***	**		Scrub and thicket important for nesting. Edges and glades for feeding.
Nightingale		**	***	**		Woodland edge, scrub and coppice. Nests in herbaceous understorey. Prefers sunny woods.
Tree pipit	**	**				Prefers open woods or clear fells. Needs isolated trees for displays.
Nightjar	**	**				Clear fells larger than 2 ha for nesting. Isolated trees for perching. Often on sandy soils.

Sources: Smart and Andrews (1985) and Fuller (1995)

....

Species Box 2.12: Great spotted woodpecker

The great spotted is the most widespread species of woodpecker in Britain, only being absent from some Scottish islands and Ireland. At 230 mm long it is intermediate in size between the green and lesser spotted woodpeckers. It is black and white in colour with large white shoulder patches and red under the tail. The males have red on the neck and the juveniles of both sexes have red caps. They are found in a wide range of woodland types but need trees showing signs of decay. The birds feed largely on invertebrates, which they obtain from the branches of trees. They have a very long tongue, barbed at the end, which is used to extract larvae from tunnels in the wood. At certain times of the year caterpillars form a large part of the diet and in the winter they will eat seeds as well. The males establish territories and attract females by drumming their bill on a hollow branch, but the females also drum. Great spotted woodpeckers make their own holes to nest in, usually in a dead branch (which is easier to hollow out) well above the ground. They may use the same nesting site in future years, but often other animals use it.
Sources: Smart and Andrews (1985); Fuller (1995); Peterson *et al.* (1983)

···

Species Box 2.13: Tawny owl

The tawny owl is the most widespread owl in Britain (but absent from Ireland) and is almost always associated with trees. It varies from red-brown to grey-brown and has a paler, black streaked breast. It is 0.38 m long with a large head and dark eyes. Tawny owls are found in a wide range of woodland types, both coniferous and deciduous and often like some open areas within the woodland. They feed on small mammals, especially wood mice and bank voles but also eat invertebrates, especially earthworms, amphibians and birds. They hunt mainly by sitting still on a perch and swooping down from it to take prey from the ground. They are very territorial and can be heard calling to establish the boundaries of their territories with characteristic calls. Nests are made in hollow trees (they do not excavate their own hollows) but they will readily nest in boxes made especially for them. The pairs stay together from one year to the next and rear up

to four young depending largely on the food supply after hatching.
Sources: Smart and Andrews (1985); Fuller (1995); Peterson *et al.* (1983)

···

Grey squirrels will also chisel out the embryos in acorns before burying them so that they are not able to grow (Evans, 1997). The red squirrel is quite a different animal, although equally adapted to arboreal life, it prefers coniferous areas and usually feeds preferentially on pine seeds. In Britain now it has a very restricted distribution.

Another native species that is totally reliant on woodland is the common dormouse (see section on coppicing). Other small mammals such as mice, voles and shrews are commonly found on the woodland floor with the wood mouse (Species Box 2.15) being perhaps the most ubiquitous. All of these species will use other habitats too.

The mammals usually less frequently seen are the bats. Bats evolved as tree-roosting animals but, due to the decline in suitable trees, some like the pipistrelle have adapted

well to living in houses. Others, like the noctule (Species Box 2.16) are still reliant on trees. A good range of tree holes in woodland is beneficial for a range of bat species including the rarest in Britain, Bechstein's bat.

Nutrient flow and decomposition

Trees, like other plants, require minerals from the soil in order to grow. Large amounts of these are released from dead plant material such as leaves, twigs and branches by the decomposition process. Decomposition is the gradual disintegration of dead organic matter (Begon *et al.*, 1996) producing carbon dioxide, water and inorganic nutrients such as nitrogen and sulphur. Microscopic bacteria, fungi and a host of animals ranging from tiny mites and springtails to larger millipedes,

Table 2.7: Bird habitat and food preferences in woodland

Species	Upper canopy (>11 m)	Low canopy (5–11 m)	Shrub layer (1–5 m)	Field layer (0.1–1 m)	Ground	Food preference
Wood pigeon	*****	*	*			Seeds, fruits, buds and leaves. Some invertebrates.
Nuthatch	**	**	*			Insects in summer from tree trunks. Also seeds from the ground in winter.
Blue tit	***	****	**	*	*	Insects and small seeds.
Treecreeper	**	***	**	**		Insects from tree trunks.
Coal tit	**	***	**	**		Insects in summer and seeds in winter.
Long-tailed tit	***	****	***	*	*	Insects and some small seeds.
Marsh tit	*	**	***	**	*	Seeds and fruits including large seeds. Also insects in breeding season.
Great tit	**	***	****	**	*	Insects and seeds including large seeds.
Goldcrest	*	*	**	*		Insects in canopy.
Robin			**	**	**	Insects and invertebrates with some fruit and seeds outside breeding season.
Wren			**	***	**	Insects close to ground.
Blackbird	*	*	**	**	***	Insects, worms and berries from ground.

Sources: Colquhoun and Morley (1943) and Fuller (1995)

Note: The number of asterisks in each column indicates the relative abundance of the bird species in that particular height zone in Bagley Wood in Oxfordshire.

woodlice and worms carry out the breakdown.

The rate at which this waste plant matter is broken down varies according to the situation at a particular site. On the whole, decomposition is slower in coniferous forests than deciduous ones and faster in warmer climates. This is illustrated by Table 2.10 which shows the mean length of time various minerals remain on the woodland floor in a variety of situations. It has been estimated that 75 per cent of the annual nutrient requirements of trees in a woodland can be met by the decomposition of one year's leaf litter fall (Davis *et al.* 1992).

Coniferous and deciduous woods generally have different types of decomposer communities. The leaves of deciduous trees are relatively easy to break down and the animals doing this tend to be large decomposers, woodlice and millipedes and worms, which mix the soil up as well. The resultant soil tends to be a mull where the organic layers are well decomposed and incorporated into the soil profile. The needles from coniferous trees are much more difficult to break down and the animals responsible tend to be the smaller mites and springtails (Species Box 2.17). These soils are called mors and generally have a deeper organic layer, with a higher proportion of un-decomposed matter at the surface. Fungi and bacteria are usually important in both systems though bacteria may be inhibited at low pH levels (M. Martin pers. com.).

Table 2.8: Characteristics of the different deer species in Britain

Species	Origin	Distribution	Notes	Food
Red deer *Cervus elaphus*	Native	Patchy. East, south, south-west and northern England and Scotland	Forest animal but adapted to open land. Herding	Grass, crops, heather, trees and shrubs
Roe deer *Capreolus capreolus*	Native (and also some introduced stock)	Widespread	Woodland and fields. Solitary	Trees, shrubs, bramble, herbs, grasses and crops
Fallow deer *Dama dama*	Introduced in eleventh century	Most of lowland England	Park deer and also mature woodland. Herding	Grass, crops, herbs, trees and shrubs
Sika deer *Cervus nippon*	Introduced 1860–1920 from E. Asia and Japan	South, north-west England and Scotland	Dense woodland or scrub. Herding	Grasses, herbs, trees, shrubs and crops
Muntjac *Muntiacus reevsei*	Introduced in twentieth century from SE China and Taiwan	Southern England and Midlands. Spreading	Dense woodland. Solitary	Shrubs, saplings, bramble, herbs, grasses and crops
Chinese water deer *Hydropotes inermis*	Introduced in twentieth century from E China and Korea	East Midlands and Norfolk	Grassland and swamp, also in woods. In small family groups	Grass and herbs. Some shrubs and crops

Source: English Nature (1997)

Table 2.9: Characteristics of woodlands with high and low deer levels

High deer densities	Low deer densities
Glades common. Woods park-like at edges	Glades rare and close over quickly. With scrubby edges
Palatable trees and shrubs tend not to regenerate	Palatable tree species commonly regenerate and may out-compete rarer species
Little sapling growth. Browse line very clear	Dense understorey to ground level
Lower part of trunks well lit. Good lichen growth	Lower trunks shaded. Little lichen growth
Ground flora short, palatable species reduced in abundance/stature	Tall ground flora. Competitive species may shade out lower growing ones
Reduced cover for small mammals and ground nesting birds	Good ground cover for small mammals and birds. May be increased predation on tree seeds and seedlings as a result
Increased ground disturbance may favour seed germination	Seedlings which germinate may not be eaten but suffer greater competition from ground flora
Increased levels of carrion and dung	Reduced levels of carrion and dung

Source: English Nature (1997)

··

Species Box 2.14: Roe deer

Roe deer stand at about 0.6–0.7 m at the shoulder, the males being slightly larger than the females. They are reddish brown in colour with paler underparts. The kids when born are paler with black flecks and white spots on the sides and along either side of the spine. Roe deer appear to have no tail, other than a tuft of hair in the females, but have a distinct white rump patch, which is obvious when the deer are alarmed. The antlers are short (less than 0.3 m) and almost vertical, with usually three tines or points on each antler and abundant tubercles (bumps) at the base. Roe deer are widespread throughout Scotland and northern England. They have a more patchy distribution in southern England occurring in western East Anglia and central southern England roughly from Surrey to Devon. Roe are found in a wide range of woodlands preferring those that are fairly open. Densities can exceed 25 per km^2 in areas of young woodland. Both males and females have their own ranges, which vary in size according to the habitat. Breeding males defend territories during the summer and they are usually the same from one year to the next. The rutting season is from mid-July to the end of August and males at this time are very aggressive. Fertilisation of the egg occurs at rutting but the embryo does not implant into the wall of the female uterus until around Christmas time when development then

proceeds. One, two or occasionally three kids are born between May and June. They are left lying alone for the first 6–8 weeks, although the mother returns regularly to feed them. Females do not usually breed until they are 14 months old. The diet of roe deer is very varied, readily taking buds and shoots of deciduous trees and shrubs. The leaves of a wide range of herbaceous plants including brambles and heather are eaten usually in preference to grasses.

Source: Corbet and Harris (1991)

··

Dead wood on the ground tends to be broken down by slightly different species of animals to the dead leaves. In woodlands most similar to the wildwood type, the quantities of dead wood on the ground can be substantial, e.g. up to 94 m^3 per hectare in parts of Białoweiża. A standard method of measuring the volume of dead wood in woodlands is outlined by Kirby (1992). Kirby *et al.* (1998) compared the amount of dead wood from a range of sites using this method. They consider values of less that 20 m^3 per hectare to be low, between 20 and 40 m^3 per hectare to be

medium and more than 40 m^3 per hectare to be high for British conditions.

Large pieces of wood are occupied by a succession of invertebrates (Figure 2.5) and the species of fungi colonising the same piece of wood may show a similar progression.

A different range of organisms break down dead wood when it falls into water (i.e. rivers and streams). Until recently the value of this type of habitat has been largely overlooked, but it is also threatened by the tidying up of rivers for salmon fishing interests (Stubbs, 1997).

··

Species Box 2.15: The wood mouse

The wood mouse is distinguished from other British mice by being dark brown on top and paler grey-white underneath, having protruding eyes, large ears and a long tail. It is extremely common and abundant throughout the British Isles and is found in a wide range of habitats, not just woodland. They seem to prefer woods with dense ground cover. During the winter months the mice may live in small groups. At the beginning of the breeding season the females develop distinct home ranges which they may defend. The young are born between March and October with two to eleven in a litter, each female having up to six litters. Only a small number of adults survive over the winter and although litters may be born during the winter months, they tend to be small and take longer to grow. Wood mice are largely nocturnal and may spend much of their time above the ground on low branches. The nests are usually below ground, made of leaves and moss and may contain caches of food. Wood mice eat a wide range of food including seed and fruits, green plants and invertebrates. They have large numbers of predators with tawny owls and weasels being the most important.

Source: Corbet and Harris (1991)

··

THE SPECIES COMPOSITION AND CLASSIFICATION OF WOODLAND

Native and introduced trees

Tree species naturally colonising after the Ice Age were severely hampered after Britain became an island, about 8,000 years ago. Thus, we have a relatively short list of 'native' species and a much longer one of trees that have been introduced by humans. Some of the introductions took place a very long time ago so that the trees feel very much part of the landscape now. Table 2.11 lists the native species and some of the introductions according to when they arrived in Britain. Only 29 broadleaves and three conifer species are considered truly native. From the ranges of the various trees in relation to climate and soil type, broad zones can be identified demonstrating the probable composition of the semi-natural woodland in Britain (Figure 2.6). The boundaries between the zones would not be as abrupt as shown here, which in any case are approximations. The ranges of trees such as beech have been discussed at length (e.g. Rackham, 1997) because they have been planted outside the region of their natural spread and have become well established there.

....

Species Box 2.16: Noctule bat

The noctule bat is the British species most dependent on trees, preferring woodpecker holes and rot holes in deciduous trees for roosting. It is one of the largest species measuring 70–82 mm long with a wing span of 330–450 mm. The noctule has rather greasy-looking fur which is ginger or reddish and slightly paler beneath. The face and wings are almost black. It feeds on large insects, which it catches and eats in flight, swooping down on them from a high level above the tree tops. The noctule flies shortly after sunset for about an hour and then just before dawn for a slightly shorter period. The males are usually solitary and hold territorial roosts during August and September. The females move between the male territories and mating takes place at this time. The semen does not fertilise the eggs until the spring and a single offspring is produced in June or July. The young are born in nursery roosts established by the females in March, although a number of different roosts are used throughout the season. The young are weaned in August and some may mate in their first year and give birth the following summer. Noctules spend the winter in trees or occasionally in buildings where both sexes aggregate. They can withstand low air temperatures and will forage throughout the winter at a reduced level.

Sources: Corbet and Harris (1991); Holmes (1997)

....

Native trees tend to have more animals and epiphytic plants associated with them than do non-native ones and are usually preferred for planting where nature conservation is an aim. However, the concept of nativeness has been questioned by Brown (1997). He points out that some of the species introduced by humans may well have eventually colonised naturally at some point. Also, that small-scale climate changes may have altered the composition of the woods anyway (for example, small-leaved lime has declined over the last 5,000 years due

Table 2.10: Rate of decomposition and mineral turnover in different woodland types

Forest type	No. of sites	Organic matter	Mean residence time of minerals in the woodland litter (years)				
			N	K	Ca	Mg	P
Boreal coniferous	3	353	230	94	149	455	324
Boreal deciduous	1	26	27.1	10.0	13.8	14.2	15.2
Temperate coniferous	13	17	17.9	2.2	5.9	12.9	15.3
Temperate deciduous	14	4.0	5.5	1.3	3.0	3.4	5.8
Tropical rain forest	4	0.4	2.0	0.7	1.5	1.1	1.6

Source: Kozlowski *et al.* (1991)

Species Box 2.17: Springtails

Springtails or Collembola are extremely common animals in woodland leaf litter, especially that under pine trees. Springtails are primitive insects, only a few millimetres long with no wings. Many are distinctive in that they have a jumping organ or furca, which is situated on the underside and is an escape mechanism enabling the animals to jump considerable distances out of the way of predators. Some species living deeper in the soil have reduced jumping abilities. Also on the underside of the body is the ventral tube, which is important in regulating fluid balance and also may help the springtail stick to slippery surfaces. Springtails feed mostly on fungi and decaying plant material and help to break down and recycle leaves. Although they are tiny, they can occur in large numbers and therefore can be very important components of some soil faunas. The female springtail lays eggs singly or in small clutches in the soil or leaf

litter. The young hatch as miniature versions of the adults but lacking the sexual characteristics. There are 5–8 instars (stages) before maturity is reached but the adults may then continue to moult.

Source: Hopkin (1997)

to a slight cooling of the climate) so the composition is unlikely to remain static. Our woods show a range of naturalness and we should consider some of our woods to be 'locally distinct' even if not composed of native species. One example is that of sycamore woods in the Pennines.

Woodland types and classification

Classification systems

British woodlands vary greatly in their character. Some have dense canopies and poor field layers, others are more open in aspect and are perhaps grazed. The soil type, pH and mois-

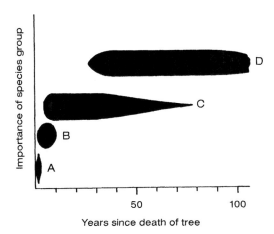

Figure 2.5: The succession of invertebrates breaking down dead wood in Fenno-Scandinavian woodlands. A: Bark feeders; B: Species living under the bark and in the surface layer of the timber; C: Wood-living species; D: Species sheltering under decaying logs. From Heliövaara and Väisänen (1984).

ture content all contribute to the composition of the wood and there are regional differences based on climatic factors too. The plates throughout this book help to illustrate this variety.

Over the years several different attempts have been made to categorise British woodlands. One of the first was published by Tansley in 1939 who was influenced by the dominant tree species but also took into account the soil type. His system provides a fairly simple classification but many woods cannot be fitted into it.

The Peterken (1993) stand type method of classifying woodlands is similar to that of Tansley in that it largely uses the tree and shrub species as the determining factors, however, it encompasses a wider range of woodland types. The Bunce (1982) plot type method uses a rather different system to provide a classification.

More recently (Rodwell, 1991) the National Vegetation Classification (NVC) has been published for woodlands. This differs from two of the previous systems mentioned (Tansley and Peterken) in making extensive use of the ground flora. The NVC also explores the relationships between the different woodland types in terms of soil type, moisture, climate (in geographical terms) and the amount of light reaching the woodland floor (see Table 2.12 for a broad overview of the woodland categories and how they relate to one another). It is worth noting that the titles of some of the communities can be a little misleading, the most obvious tree species in a wood may not be reflected in the name of the community. For example most W8 (ash – field maple) might be considered oak woods under other systems (Plate 2.5). Table 2.13 provides a rough comparison of the different classification systems. The NVC is the classification system most widely used in Britain at the moment, especially in sites of nature conservation interest, and Table 2.14 gives more details of the composition of the various communities and sub-communities (see Rodwell (1991) and Whitbread and Kirby (1992)). The NVC system has also been used to plan the species composition of new woodlands (Rodwell and Patterson, 1994).

The Forestry Commission is developing a rather different method. The ecological site classification (ESC) looks at the soil nutrient levels and soil moisture regimes. Then, the most appropriate tree species for planting can be identified (Malcolm, 1997).

Woodland zones

The species composition of woods varies on an altitudinal basis. This is more pronounced in, for example, the Andes, where the progression from riverine gallery forest through tropical rain forest to montane cloud forest and then

Table 2.11: The native trees of Britain and those introduced, with approximate date of introduction

Native species (in approx. order of arrival)	Introduced species				
	before 1600	1600–1700	1700–1800	1800–1900	After 1900
Juniper	Walnut	Black walnut	Corsican pine	Douglas fir	Dawn redwood
Downy birch	Cornish elm	Cedar of Lebanon	Cricket bat willow	Eucalyptus	Hybrid wingnut
Silver birch	English elm	Common lime	Grey alder	Giant fir	Rauli
Aspen	Grey poplar	Cork oak	Lombardy poplar	Hybrid black poplars	Roble
Scots pine	Evergreen oak	Dutch elm	Red oak	Italian alder	
Bay willow	Laburnum	European larch	Turkey oak	Japanese larch	
Alder	Maritime pine	False acacia	Monkey puzzle	Lodgepole pine	
Hazel	Norway spruce	Horse chestnut		Noble fir	
Small-leaved lime	Stone pine	London plane		Robusta poplar	
Bird cherry	Swedish whitebeam	Norway maple		Serbian spruce	
Goat willow	Sweet chestnut	Red maple		Sitka spruce	
Wych elm	Sycamore	Scarlet oak		Southern Beeches	
Rowan	White poplar	Swamp cypress		Wellingtonia	
Sessile oak	Wild pear	Tulip tree		Western balsam poplar	
Ash				Western hemlock-spruce	
Holly				Western red cedar	
Pedunculate oak					
Hawthorn					
Crack willow					
Black poplar					
Yew					
Whitebeam					
Midland hawthorn					
Crab apple					
Wild cherry					
White willow					
Field maple					
Wild service tree					
Large-leaved lime					
Beech					
Hornbeam					

Sources: Hart (1991) and Lane and Tait (1990)

alpine-type vegetation can, in some places still be seen. Although Britain does not have particularly high mountains and the lowland areas are very disturbed by cultivation and development, it is worth noting that some remnants of the extreme zones can still be distinguished.

At low elevations, along water courses there were riparian and flood plain forests. These types of woodland are extremely rare in Brit-

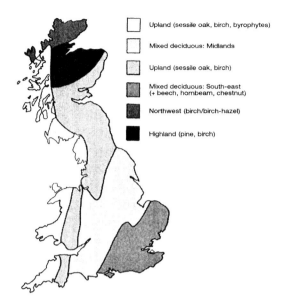

Upland (sessile oak, birch, byrophytes)

Mixed deciduous: Midlands

Upland (sessile oak, birch)

Mixed deciduous: South-east
(+ beech, hornbeam, chestnut)

Northwest (birch/birch-hazel)

Highland (pine, birch)

Figure 2.6: The zones of semi-natural woodland in Britain. From Fuller (1995).

ain now (Plate 3), partly because of the high degree of control we apply to river systems. These woods are generally characterised by episodic flooding and Peterken and Hughes (1995) recognise four types which vary in the moistness of the soil:

1 pioneer fast-growing poplar and willow woodland, frequently flooded;
2 alder fen carr to acid mires with grey willow;
3 mixed broadleaved elm, oak, ash (Species Box 2.18) and alder, sometimes flooded in winter, rarely in summer;
4 oak, hornbeam, lime woodland on flood-plain margins on the highest ground and rarely flooded.

They note that although there are small fragments of wet woodland in Britain, most

Table 2.12: A general outline of the National Vegetation Classification

			Willow	Birch	Alder
Wet soils			W1 (Grey willow – marsh bed-straw) W2 (Grey willow – common reed) W3 (Bay willow – bottle sedge)	W4 (Downy birch – purple moor grass)	W5 (Alder – greater tussock sedge) W6 (Alder – nettle) W7 (Alder – ash–yellow pimpernel)
Dry soils			*Alkaline*	*Neutral*	*Acid*
Upland	Cold north	Montane	W20 (Downy willow – great Wood-rush)	W19 (Juniper)	W18 (Pine)
	Cool and wet, north and west	Sub-montane	W9 (Ash – whitebeam)	W11 (Oak – birch – wood sorrel)	W17 (Oak – birch – *Dicranum majus*)
Lowland	Warm and dry, south and east		W8 (Ash – field maple)	W10 (Oak – bracken – bramble)	W16 (Oak – birch – wavy hair grass)
	Zone of natural beech		W12 (Beech – dog's mercury)	W14 (Beech – bramble)	W15 (Beech – wavy hair grass)
	Local in southern Britain		W13 (Yew)		

Sources: Rodwell (1991) and Martin pers. com.

Plate 2.5: Oak/ hazel woodland, Wetmoor Wood, Gloucestershire

are secondary in nature and correspond to 2 above (see also Plate 2.6).

At the other altitudinal extreme, montane woods have a very different character. On the tops of mountains (varying between 300–500 m in Britain) the conditions are too harsh for tree growth so a tree line is formed. The tree species found at highest altitudes tend to be coniferous with deciduous trees only surviving to lower levels. Often the trees are reduced in stature, such as the Scots pine of the montane scrub in Scotland (Edwards, 1997).

CHANGES IN WOODLANDS

Partly as a result of the storms in 1987 and 1990 ecologists are increasingly realising that disturbances are a natural part of woodland ecosystems in Britain. Severe disturbances may only occur sporadically, with long intervals between them but they can have significant effects on the woodland. Brown (1997) notes that because trees are long-lived organisms, disturbance effects may well occur within the

life of the tree. A few types of natural change will be discussed here and some longer-term human-induced changes. Peterken (1996) discusses disturbance in woodland in more detail.

Natural changes

Peterken (1996) lists a wide range of natural disturbances including wind, fire, ice and snow, drought and landslips. Most British trees are not very combustible so that fire is not so much part of the natural system as it is in other parts of the world (e.g. Australia) where some trees require fire for the seeds to germinate. Occasionally it can have a significant effect, e.g. in native pinewoods or, in a much more man-influenced landscape, Ashtead Common in Surrey where fire spread in bracken litter and killed many old oak pollards, which being hollow burnt more easily. Fire can also be a significant commercial problem in coniferous forests.

Ice and snow damage causes small-scale tree loss in parts of Britain and is a more important

Table 2.13: The classification of woodland types using some common methods

NVC code	NVC description (Rodwell, 1991)	Peterken (1993) stand type equivalent	Bunce (1982) plot type equivalent	Tansley (1911, 1939) equivalent
W1	Grey willow – common marsh bedstraw		32	
W2	Grey willow – downy birch – common reed			Fen carr
W3	Bay willow – bottle sedge			
W4	Downy willow – purple moor grass			
W5	Alder – greater tussock sedge	Alder 7Ba and 7Bb	14	Betulo-Alnetum
W6	Alder – nettles	Alder 7Aa and 7Ab		
W7	Alder – ash – yellow pimpernel	Alder 7Aa, 7Ab, 7Bc, 7D & 7Eb	12, 13, 14 & 16	Quercetum roburis
W8	Ash – field maple – dog's mercury	Ash-wych elm	1, 2, 3, 4, 5, 6, 7, 8 & 12	
		Ash-maple		
		Hazel-ash		
		Ash-lime		
		Hornbeam		
		Suckering elm		
W9	Ash – rowan – dog's mercury	Ash-wych elm 1Ab & 1D		Upland ash
		Hazel-ash 3C		
		Alder 7D		
		Birch 12B		
W10	Pedunculate oak – bracken – bramble	Hazel-ash	9, 17,19, 20 & 24	Quercetum roburis
		Ash-lime		Quercetum arenosum roburis et sessiliflorae
		Oak-lime		Quercetum petraeae sessiliflorae
		Birch-oak		
		Hornbeam		
W11	Sessile oak – downy birch – wood sorrel	Hazel-ash	21, 22, 23, 26 & 29	Highland birch woods
		Oak-lime		Highland oak woods
		Birch-oak		
		Birch		

Table 2.13: contd

NVC code	NVC description (Rodwell, 1991)	Peterken (1993) stand type equivalent	Bunce (1982) plot type equivalent	Tansley (1911, 1939) equivalent
W12	Beech – dog's mercury	Beech 8Cb & 8Cc	1 & 8	Fagetum sylvaticae calcicolum Fagetum rubosum
W13	Yew			Yew woods
W14	Beech – bramble	Beech 8D	17 & 20	Fagetum rubosum
W15	Beech – wavy hair grass	Beech 8A & 8B	17	Fagetum arenicolum/ericetosum
W16	Oak spp. – Birch spp. Wavy hair grass	Birch-oak Birch	18	Quercetum arenosum roburis et sessiliflorae Quercetum ericetosum
W17	Sessile oak – downy birch – *Dicranum majus*	Birch-oak Birch	18	Quercetum roboris Quercetum petraeae/sessiliflorae
W18	Scots pine – *Hylocomium splendens*	Pine	28	Pinetum sylvestris Highland pine forest
W19	Juniper – wood sorrel			Birch–juniper

Source: Rodwell (1991)

Notes: Scrub categories not included.
Note that the various different classification systems are not directly compatible. To classify a woodland using any of these systems it is necessary to use more complete descriptions than those presented here.

Table 2.14: Characteristics of the National Vegetation Classification categories

NVC code	Title	Additional species in tree and shrub layer*	Additional species in field layer and ground layer*	Comments
W1	**Grey willow – marsh bedstraw**	Downy birch, alder	Brambles, ivy, Yorkshire fog, velvet bent	
W2	**Grey willow – downy birch – common reed**		Very variable	
W2a	Alder – meadow sweet			pH 6.5–7.5, fluctuating water table
W2b	Sphagnum			pH lower, peat above water table
W3	**Bay willow – bottle sedge**	Grey willow	Meadow sweet, horsetails, common and marsh valarian	Canopy low, field layer tall
W4	**Downy birch – purple moor grass**	Grey willow (as shrub layer)	Sphagnum	Open woodland
W4a	Broad buckler fern – bramble			Less Sphagnum, some nutrient enrichment
W4b	Sphagnum			Damp wet peat
W4c	Soft rush		Sphagnum and grasses	
W5	**Alder – greater tussock sedge**	Grey willow	Large sedges	If water table rises may have dead trees
W5a	Common reed	Alder dominant	Tall herbs and sedges	
W5b	Yellow loosestrife	Shrubs: buckthorn, guelder rose	Ferns	
W5c	Opposite-leaved golden saxifrage		Smaller herbs and mosses	Seepages and springs
W6	**Alder – nettle**	Grey willow, hawthorn, elder		
W6a	Typical	Alder, grey willow	Nettles, common cleavers	
W6b	Crack willow		Common cleavers	Fallen branches, open mud
W6c	Osier – almond willow		Nettles	
W6d	Elder		Enchanter's nightshade, dog's mercury, wood avens	
W6e	Downy birch		Brambles, honey suckle, broad buckler fern	

Table 2.14: contd

NVC code	Title	Additional species in tree and shrub layer*	Additional species in field layer and ground layer*	Comments
W7	**Alder – ash – yellow pimpernel**	Grey willow, downy birch	Low growing herbs, creeping buttercup, creeping soft grass	
W7a	Nettle	Ash, alder		Free draining damp soils
W7b	Remote sedge – marsh thistle		Diverse, tall herbs, sedges and horsetails	Unstable, wet soils with water seeping to the surface
W7c	Tufted hair grass	Alder, ash, downy birch		Impeded drainage
W8	**Ash – field maple – dog's mercury**		Dog's mercury	Calcareous soils
W8a	Primrose – ground ivy	Ash, oak, field maple, hazel, lime, hornbeam		
W8b	Wood anemone	As 8a, hornbeam, aspen	Bluebell	Damp, heavy clay soils
W8c	Tufted hair grass	As 8a	Less dog's mercury and bluebell	Wet trampled soils
W8d	Ivy	Dense	Poorer in species	Often secondary woodland
W8e	Herb robert	Ash, birch, cherry, yew, whitebeam, hazel, elder, hawthorn	Dog's mercury, common cleavers, herb robert	
W8f	Ransoms	As 8e	Ransoms followed by dog's mercury	
W8g	Wood sage	Very varied	Very rich	
W9	**Ash – rowan – dog's mercury**	Hazel	Bluebell, ferns, wood sorrel, grasses good mosses	
W9a	Typical	Well developed	Short	More open community
W9b	Marsh hawksbeard	Less well developed	Tall and rich	
W10	**Pedunculate oak – bracken – bramble**	Sessile oak, silver birch, ash, sycamore, lime, hornbeam	Bluebell, wood anemone, bramble, honeysuckle, bracken	Slightly acidic soils
W10a	Typical	Hazel	Bluebell	Dry oak birch woods
W10b	Wood anemone	Chestnut	Wood anemone	Heavy soils, wet in winter
W10c	Ivy	Holly	Ivy	
W10d	Yorkshire fog	Oak, conifers	Ruderals	Sometimes in conifer plantations
W10e	Sycamore – wood sorrel	Hazel	Grasses, ferns, mosses	Most common in north west

Species Box 2.18: Ash

Ash is one of the largest deciduous trees in Europe and can grow to 40 m in height. The leaves consist of six to twelve pairs of oval leaflets on a stem with a single one at the end. The paired black buds and the smooth grey bark are obvious features of the tree in winter. The flowers emerge in the spring before the leaves and can be male, female or hermaphrodite (in varying combinations on different trees). The female flowers develop into green seeds (with a single wing) which hang in bunches and are often known as keys. They may persist on the tree for a considerable time before being dispersed by the wind. Ash regenerates well in Britain (although there are heavy seedling losses due to browsing) and is a common tree especially on moist, rich soils. It is not very shade-tolerant and so is easily shaded out by other trees such as oak and beech. The wood of ash trees is very strong and useful for a wide range of products including tool handles and sports equipment.

Sources: Grime *et al.* (1988); White (1995)

factor further north, the weight of the snow causing trees and branches to bend and snap. The shape of mature coniferous trees means that snow is shed easily (Peterken, 1996) so often it is the deciduous trees which are more vulnerable. Landslips caused by rain loosening the ground can cause problems in certain areas depending on the topography and geology.

In recent years it is wind, drought and disease which have had the biggest effect on British woodlands.

Wind

Wind damage to trees can take many forms but the two most frequently seen are either the

Plate 2.6: Alder carr at Thelnetham Fen, Suffolk

stem of the tree snapping or the whole tree being uprooted because the root-soil holding strength is less than the wind strength. The latter situation results in large vertical root plates rising out of the ground. Sometimes the trees continue growing, but usually too many roots are severed, and almost always in the case of conifers, they do not grow again. Conifers (and broadleaves during the summer months) tend to be more susceptible to wind damage and the situation is usually worse if the ground is wet. Trees with small root systems due to shallow soil are also more susceptible, though this can sometimes result in stunted trees which may then be able to withstand the wind better. (Kozlowski *et al.*, 1991 gives more details of how the wind affects trees).

The 1987 storm in Britain is an interesting example of the effects of wind on trees. The event happened in October, just after rain, so many broadleaves were affected. Approximately half the trees blown down were broadleaves (mainly beech) and half were coniferous (mainly pine) (HMSO, 1988). Both of the species mentioned here are shallow-rooted. Much of the damage was in plantations, these presented an even resistance and solid face to the wind. At the time there was a strong emphasis on clearing and replanting and substantial sums of money were made available to do this work through Task Force Trees (Countryside Commission 1991, 1993a). However, as both Rackham (1990) and the HMSO (1988) report point out, the nature conservation aspect of many woods was not as badly affected as first thought. The storm increased structural diversity and in many areas good natural regeneration has resulted. The amount of wood on the ground was beneficial to many organisms and in many places it was the younger trees which suffered most, with older ones perhaps shedding some branches but not being lost completely. Some of these issues were discussed in a symposium in 1994 on the ecological effects of the storm. Some useful points have been made as a result of this storm. In the interests of nature conservation this event was not an unmitigated disaster, there is more to a wood than the trees (HMSO, 1988). As Rackham (1990) points out, 'a horizontal tree – alive or dead is at least as good a habitat as an upright one'.

Drought

Drought situations occur when there is not enough water in the ground to supply the trees and the evapo-transpiration rates from the leaves are high. The evaporation of soil water is an important loss of potential water for the trees, as well as lack of rain. A similar situation occurs during the winter months (see section on evergreen and deciduous habits, p. 18) but here we are concerned with drought during the summer. Trees growing in parts of the world where hot dry summers are the norm (e.g. the Mediterranean) are able to withstand dry conditions to a greater extent than can British species. A drought situation can be made worse for a tree if surrounding vegetation is better at competing for the available water. This can be a problem for example, where rhododendrons are growing close to a tree (pers. obs.). The stress caused by drought has many effects on trees (Kozlowski *et al.*, 1991) but the most obvious is usually wilting, browning and eventual loss of the leaves.

Species have different degrees of tolerance to drought situations. For example in the dry summers in the early 1990s in Burnham Beeches (Buckinghamshire) silver birch were the first trees to show signs of stress, then beech, while oak rarely appeared stressed. This may be at least partly due to their root structure, oak having a deep tap root.

Recent consecutive years of dry summers have had an appreciable impact on British trees and woodlands. Power (1994) looked at the effect of drought on beech trees at a range of sites in southern England by recording the twig growth each year. She found a substantial reduction in growth during 1976 and 1977, years of severe drought. In subsequent years trees which were regarded as healthy at the start of the study made a full recovery whereas those which were unhealthy failed to do so and many continued to show reduced growth rates for many years afterwards.

Peterken and Mountford (1996) recorded the impact of the 1976 drought on Lady Park Wood (Gwent). Here many mature beech trees were killed or severely damaged. Fifteen years later, damaged trees were still dying from drought-induced stress and this has had a marked impact on the structure of the wood.

Tree diseases

In the course of this century, tree disease has had a significant effect on the trees and woods of Britain. Perhaps the best-known example is Dutch elm disease, which is caused by a fungus *Ceratocystis ulmi* spread between trees by two beetles of the genus *Scolytus*. Figure 2.7 illustrates the life history of Dutch elm disease. Two recent epidemics of this disease have occurred, the first peaked in 1936 when around 10–20 per cent of the elms in southern England were killed, but many infected trees recovered. From the late 1960s the second outbreak has taken place caused by an aggressive strain of the fungus probably imported from Canada in the wood of rock elm. By 1977, 50 per cent of the original 23 million elms in southern Britain were dead and the disease had spread up into Scotland (Figure 2.8 illustrates this spread). English elm was a characteristic species of hedgerows and open land and a very prominent feature of the landscape but elm species were also a very important component of small woods. The loss of elms was so great that it inevitably had an impact on the woods. For example, in Oxfordshire over 90 per cent of the elms were lost and in many copses it had been the dominant tree species. In Bristol over 20,000 elms died (including over 95 per cent of the English elm) and some small woods, of around 2 hectares, lost all their trees (Clouston and Stansfield 1979).

Also of concern is oak die back. This disease has occurred at various times during this century, notably in 1927 and 1989. It seems to be a 'complex disease' or 'dieback decline' (Gibbs and Greig, 1997) where a range of different

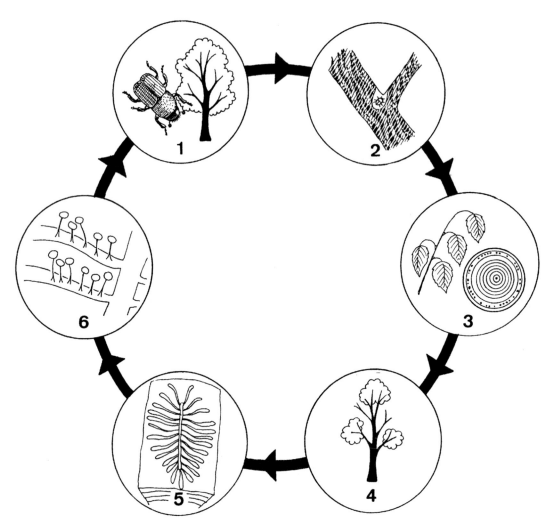

Figure 2.7: The life cycle of Dutch elm disease. From Clouston and Stansfield (1979). 1. The adult beetles emerge in the spring and summer from the bark of dead and dying trees which are infected with the fungus. The adults carry fungal spores on them. 2. The adult beetles feed in the twig crotches of healthy elms. In doing so they introduce fungal spores into the tree. 3. The infected branches of the trees wilt and the twigs show dark spots or streaks in the wood. 4. The trees which are weakened by the disease become breeding sites for beetles. 5. The beetle larvae tunnel under the bark producing the galleries. 6. The fungus grows in the insect galleries and produces spores.

factors cause the dieback or death of the tree. It occurs largely in pedunculate oak. The outbreak in 1927 was thought to be due to extensive foliage damage caused by the oak roller moth, followed by oak mildew and honey fungus. In subsequent occurrences in Britain and other European countries factors are thought to have included drought (Britain 1989 and 1990), cold winters (Germany 1985 and 1987), fungal infection by species of *Collybia* and possibly some connection with the oak jewel beetle (Species Box 2.19) (Britain 1997).

Figure 2.8: The spread of Dutch elm disease in Britain between 1971 and 1977. From Clouston and Stansfield (1979). The stippled area indicates that part of Britain severely affected in 1971.

However the beetle may just favour the conditions in the tree after it has started to decline, (Alexander pers. comm.); air pollution and other species of fungus cannot be ruled out (Gibbs and Greig, 1997).

Other potential problem diseases include beech bark disease and fungal infection by *Phytophora* in alders. Honey fungus has long been considered to be a threat to British trees. In fact there are now thought to be seven European species and that most commonly found in broadleaved woodlands (*Armellaria*

gallica) is only weakly pathogenic (Gibbs and Greig, 1997). Honey fungus seems to be more of a problem in isolated and ornamental trees than more natural woods. Although it may cause problems in conifer plantations in the first few years after planting, it also occurs in the bases of trees which are weakened by other agents (Hibberd, 1991). The fungus *Heterobasidion annosum*, however, is a serious forestry pest species. It gains access to the roots of live trees by underground root connections between live healthy trees and dead stumps where the fungal spores have been deposited by the wind. *Heterobasidion* rots away the wood at the bases of the trees (producing small brackets at ground level) making them very susceptible to wind damage. It may also kill some conifer species. An effective method of control is to paint the cut surfaces of stumps with another fungus species (*Peniophora gigantea*) which out-competes the *Heterobasidion* (Hibberd, 1991).

Human-induced changes

Fragmentation

The extensive felling of the past has resulted in the remaining woods being reduced in size and scattered unevenly across the landscape. This fragmentation has implications for the species living in these habitats, since as they get smaller they may not be able to support viable populations and if they are widely scattered, organisms may not be able to move between them.

Zuidema *et al.* (1996) outlined four factors that may cause the loss of forest biodiversity as fragmentation occurs:

1 sample effect (patchily distributed species may or may not be present in small parts of the forest);

Species Box 2.19: Oak jewel beetle

The oak jewel beetle is metallic green-blue in colour and is a representative of a family of beetles more common in the tropics. Often the eggs are laid by the females in trees that seem not to be in the best of health. The larvae tunnel in and under the bark of oak trees, sometimes the tunnelling larvae may girdle branches causing them to die back. The adults emerge from characteristic D-shaped holes in June. The oak jewel beetle was listed as a Red Data Book species in 1987 and was considered to be confined to old woods with oak trees. It seems to have become more widespread in recent years.

Sources: Gibbs and Greig (1997); Shirt (1987)

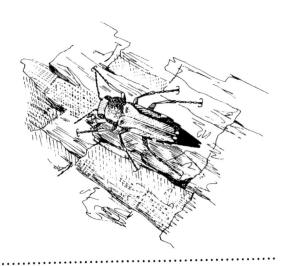

2 forest isolation (the further apart suitable fragments are, the more difficult species will find it to move between them and recolonise);

3 reduced forest size (some species need large areas in order to survive);

4 newly created edges (microclimate is altered, sometimes to a considerable distance into the wood and other non-forest species can colonise).

Figure 2.9 illustrates how this process might work. The degree to which different groups respond to fragmentation depends upon their characteristics. Birds are mobile and can move to new sites relatively easily but they may need large areas of woodland to support a viable population. In contrast, plants can persist in smaller woods but are sometimes less able to recolonise new areas. It is important to note that fragmentation may increase the overall biodiversity. However, species coming in will not be true forest species but those characteristic of more open or disturbed habitats.

Gibson (1988) studied a range of different woods in Oxfordshire and discovered that for ancient woodlands there was a good relationship between the number of ancient woodland indicator plant species and the area of the habitat patch, i.e. there were more species in larger areas. More recently wooded sites had fewer ancient woodland species than expected for their size.

Insects seem particularly susceptible to forest fragmentation in the tropics (Didham *et al.*, 1996) where not only the abundance and diversity of insects are affected but changes in the interactions between insects and other organisms also occur.

Acid rain

The decline in tree health due to pollution has been a cause of concern in central Europe (and more recently in Britain) since the 1970s. Rainfall is naturally acidic, ranging from pH 4.5 to over pH 5.6 (Innes, 1987). It becomes more acidic when pollutants such as sulphur dioxide and nitric oxide from burning or car exhausts react with other substances in the air to form

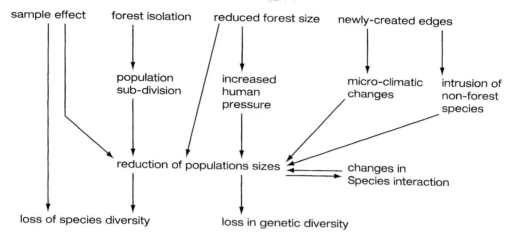

FOREST FRAGMENTATION

Loss of forest biodiversity

Figure 2.9: A theoretical scheme indicating the impacts of forest fragmentation. From Zuidema *et al.* (1996).

sulphuric acid and nitric acid. As well as falling as rain, this increased acidity is deposited during mist and fog and also as dry deposition.

The structure of trees encourages more deposition of pollutants relative to areas without trees (DoE, 1993) so that soils beneath woodlands are likely to become even more acidic. This has consequences for streams and rivers running through wooded areas and may also be detrimental to the trees due to nutrient imbalances and detrimental effects on mycorrhizal fungi. The increased acidity of the rain directly falling onto trees is considered likely to contribute to tree damage and can also alter the sensitivity of the trees to other factors, e.g. frost damage, wind, drought and pests.

The Forestry Commission tree health surveys started in 1984 but only since 1987 have they been comparable with European work. Over thirty different measurements of crown condition are recorded but the surveys only cover visible damage (DoE, 1993). Some of the broadscale effects of tree decline are shown in Table 2.15. Results indicate that trees in the UK are generally poorer in health than those in other

European countries and have declined in recent years but it is difficult to pinpoint the exact reasons for this. The health of beech in southern England in particular seems to have declined.

Global warming

Carbon dioxide in the atmosphere plays an important part in global temperature regulation by absorbing the infra-red part of the sun's radiant energy (i.e. the heat rays). The more carbon dioxide there is in the atmosphere, the more heat is absorbed and the higher the temperature of the earth. Plants need carbon dioxide to grow and trees absorb a large amount during their long life. When trees die, carbon dioxide is released as the wood decomposes, it is then taken up by other trees which are still growing. However, as forests are cut down and burnt, more carbon dioxide is released than is taken up by the relatively small number of trees planted. It has been estimated that the total amount of carbon held by forests is 1.2 billion tonnes (out of a total 2 billion held in all terrestrial plants and soils).

Table 2.15: The features of decline in different tree species

Norway spruce	Silver fir	Beech	Ash	Oak
Chlorosis and yellowing of older leaves	Chlorosis and yellowing of older leaves	Leaf yellowing in summer, development of marginal necrosis	Delay in shoot initiation	Crown thinning
Reddening of older needles	Casting of older needles (starting from base and inner parts of crown)	Loss of green leaves from June	Failure of leaves to turn green	Branch decay
Casting of older needles (starting from base and inner parts of crown)	Reduction in height increment	Development of serrulate leaf edges	Crown thinning	Leaves develop in clusters and production of short shoots
Hanging down of secondary branches (tinsel syndrome)	Reduction in fine root system and mycorrhizal regeneration capacity	Production of small leaves	Development of clusters of leaves	Premature leaf fall without autumn colours
Reduction in fine root system and mycorrhizal regeneration capacity		Formation both epicormic and short shoots	Reduction in shoot length	Black or red spots on the bark, secretion of liquid, deep cracking and premature necrosis
		Disruption of natural regeneration close to trunks	Development of lateral branches	Leaves emerge a luminous yellow colour showing the veins through clearly (pedunculate oak)
		Crown thinning		Leaf edges less indented (pedunculate oak)
		Branch dieback		Leaf curl towards the end of summer (pedunculate oak)
		Bark necrosis		
		Leaf curl		
		Abnormal branching patterns		

Source: Ling and Ashmore (1987)

Notes: Necrosis – dying; chlorosis – yellowing; serrulate – crinkly edges.

Of this a half is in boreal forests, one third in tropical forests and one seventh in temperate forests (Dixon *et al.*, 1994). The destruction of forests therefore makes a significant contribution to global warming. The average temperature of the earth has increased by about 0.5°C in the past 80 years and is expected to increase by 0.5°C in the next decade (DoE, 1996, cited in Brown, 1997). While it is generally agreed that the temperature is likely to rise, how this will actually affect the British climate is less certain so the consequent effects on the plants and animals are even more difficult to predict. It is likely that there could be quite serious effects on the woodlands. Increases in temperature may encourage trees such as beech, small-leaved lime and field maple to expand their range northwards (Fuller, 1995), however, spring-flowering herbaceous species may decline. The fauna will undoubtedly be influenced too. A recent alternative scenario is that the Gulf Stream will shift or stop, and this would be likely to cause a drop in temperature rather than a rise.

THE FUTURE

It seems inevitable that British woodlands will change in future years. While the extensive planting of new woods can contribute to the reduction of global warming, a country the size of Britain cannot have more than a minor influence on world conditions. Some respondents to the Forestry Commission document on woodland creation did address the question of how planting could take account of possible environmental changes, their suggestions were as follows (Forestry Commission/Countryside Commission 1997):

1 Continue research and monitoring and update current standards.
2 Plant a range of species and avoid monocultures.
3 Plant native species most suited to the location and encourage natural regeneration.
4 Use robust species and avoid complicated mixtures.
5 Use entirely new species.
6 Plant species in the middle of their ranges.
7 Avoid areas susceptible to natural disasters.
8 Link woodlands to allow migration of species.
9 Plant large not small woodlands.
10 Assess the site carefully.
11 Develop new planting and management techniques especially during the establishment phase when the plants are particularly susceptible to drought.

3

MANAGEMENT AND CONSERVATION

●

WHY MANAGE WOODLANDS?

The aim of managing woods in medieval times was entirely for their products, many of which were essential to everyday life. It is probable that almost all trees in the landscape were 'working trees' (Green, 1994). Woods yielded timber (from trees over two feet in girth) for building houses, ships, etc. and underwood, smaller diameter wood from coppice, pollards, suckers and the branches of the timber trees for tools, fuel, etc. A wide range of historical and current uses can be seen in Tables 1.3 and 3.1.

While we are less reliant on traditional wood products today, it is still an essential commodity and one of our few natural renewable resources. However, further reasons for management are currently evident such as conservation and public amenity. Table 3.2 shows the results of the Oxfordshire Woodland Project Survey. It lists the objectives of the owners but does not mention whether the woodland is actively managed for the objectives. Nevertheless it is interesting to note that fewer than 1 in 6 owners were not expecting to obtain direct economic returns for their woodland (Forestry Commission/Countryside Commission 1997).

Emphasis in recent years has been directed towards conservation and 'biodiversity' to the extent that these are the main reasons for managing many woods today. However, the ideal method of management for conservation is not always clear. Groups of organisms such as birds, mammals, flowering plants and butterflies have received a disproportionate amount of attention in relation to the number of species involved. These groups tend to be very visual ones, are often easier to identify and frequently the ecological requirements of the species are relatively well known. As an example, there are approximately forty species of butterfly which are considered to be associated with woodlands but nearer to 1,000 flies (Marren, 1990). Despite this, management techniques such as coppicing are frequently implemented primarily to benefit groups such as butterflies.

WOODLAND MANAGEMENT TECHNIQUES

Rackham (1988) gives three traditional methods of managing trees and woods:

1 as woodland producing timber (from maidens or standards, i.e. uncut trees) and underwood, mostly managed as coppices;
2 as wood-pasture where the trees were pollarded or parts of the area fenced temporarily to keep out the stock;
3 non-woodland trees, e.g. hedges or trees in fields which were managed as coppice, pollards or timber trees.

To this a fourth method can be added from 1600 onwards (and in some places possibly

Table 3.1: The past and current uses of wood from some deciduous trees

Species	Uses
Alder	Clog soles, tool handles, bobbins, hat blocks, toys, peeled veneer, scaffold poles, faggots, furniture, gun powder, brush backs, river revets, small boxes, instrument cases, underwater piles, levers, cross-bows
Apple	Tool handles, blocks for woodcuts, golf clubs, mallet heads, turnery
Ash	Potters crates, tool handles, felloes for wheels, pole turnery, hoops for sails, masts for yachts, sports equipment (e.g. hockey sticks, gymnasium equipment, billiard cues and cricket stumps), ladders, barrel hoops, wattle (for wattle and daub), butchers meat trays, snow shoes, spear shafts, lances, arrows, barrels, paddles, chair backs, fence stakes, walking sticks, cart shafts, crutches, car and plane frames, boats, hop poles, building timber, handbarrows, tent pegs, horse jumps and police truncheons
Aspen	Wood pumps, ties, fruit boxes, besom brush handles
Birch	Horse jumps, handleless besoms for vinegar vats, cotton reels, tannin (from bark), poles, faggots, brooms and brushes, furniture, plywood, small tool handles, turned products
Elm	Turnery, pit wood, firewood, beetle (mallet) heads, water wheel paddles
Hazel	Hurdles, thatching spars, barrel hoops, trackway drainage (as bundles), tool handles, pottery crates, clothes props, ethering rods for hedgelaying, pergolas, cotton reels, packing sticks for fruit baskets, pheasant traps, fishing rods, rabbit and deer snares, shepherd crooks, ox-yoke bows, hay creels, straw spars for bee keepers, wattle for wattle and daub walls, clothes pegs, pea sticks, upper rims of coracles, fish traps, walking sticks, faggots for bakers ovens and clay kilns, sheep cages, summer houses, water divining rods, simple ropes, catapults, bows and arrows, bird traps, sheep couplings, hedging stakes
Hornbeam	Engineer's wood, roller brushes, shoe lasts, piano keys, plane stocks, cogs, pulleys, screws, skittles, balls, brush backs, golf club heads, mallets, firewood, pulpwood
Lime	Carving wood, hat blocks, turnery wood, small boxes, veneer for chip baskets, spoons, ditching shovels, tool handles, pattern making, joinery, frames for bee hives, hop poles, ropes (from the outer bark or bast)
Maples	Drinking bowls, egg cups, musical instruments, chopping boards, writing tables, weavers shuttles, furniture and kitchen surfaces, platters and bread boards, rolling pins, ladles, butter prints, fence stakes, spigots, spoons, ox yolks, clothes rollers, decorative veneers, turning and carving, violins, boxes, chemists pestles, firewood
Oak	Wind and water mills, hubs of cart wheels, fencing, tanning (bark), ship building, roofing shingles, wood carving, hurdles, knife handles, cross bars on telegraph poles, gate posts, timber frames for buildings, barrels and tubs, gates, furniture, mining props, ladder rungs, rail wagons, turnery, walking sticks, buckets, decorative work, pegs, dowels and wedges
Sweet chestnut	Walking sticks, hop poles, fencing, rails, hurdles, ladder rungs, stakes, tent pegs, fruit props, gate posts, window sills, garden seats, clothes props, hoops
Willow	Basket making, barrel hoops, thatching spars for trugs, fuel for ceramic kilns, seats in coracles, chip baskets, artificial limbs, polo balls, broom fibre, tally sticks, scythe snaiths, wire haulage drums, boards for whetting knives, packing cases for furniture, pegs, clog soles, tool handles, hurdles, clay spades, brake blocks and flooring of carts, blades of boat paddles and water mills, yokes for milkmaids, milk pails, veneers for decorative boxes, basketry, trug baskets, pea sticks

Sources: Hampshire County Council (1991); BTCV (1997)

Table 3.2: The objectives of 256 woodland owners in Oxfordshire

Objectives of owners	Main objectives (%)*	An objective (%)
Conservation	30	84
Landscape	28	90
Timber production	9	62
Sporting	9	34
Recreation (private)	5	36
Recreation (public)	4	0.5
Shelter	6	27
Screening	2	6
Education	1	8

Source: Forestry Commission/Countryside Commission (1997). Derived from the Oxfordshire Woodland Project.

Note:
* This was just one of many questions asked. Not all respondents answered so the numbers do not add up to 100%.

earlier), that of plantations of trees managed for timber alone.

Of the four management techniques mentioned here, three will be described below (omitting non-woodland trees) together with their implications for nature conservation. In addition, other management issues will be discussed such as grazing and recreation.

COPPICE

Coppicing is one of the classic woodland management techniques. Although it is not as widely practised now as it was in the past, the decision to coppice or not is a frequently debated subject and one worthy of considerable discussion here.

The word coppice comes from the French word *couper* meaning to cut. An area of coppice is where the trees or shrubs have been cut at ground level and subsequently many shoots have grown up. The cut stump is called the stool and the shoots are sometimes known as the spring. A coppiced woodland is divided up into blocks with one (or more) being cut each year on a rotational basis. The cut area is called by various different regional names. Some of the common ones are coupe, panel, hagg, cant, fell, sale and burrow. A range of different tree species can be cut in this way. Hazel (Species Box 3.1) was probably the most commonly cut in the past but ash, oak, lime, elm and even beech were also coppiced. The greatest area of coppice cut regularly now is sweet chestnut. Most trees grow very quickly for the first few seasons after cutting, for example, sallow can reach 3.3 m and oak 2.1 m in the first season (Rackham, 1990). The length of time between cuts depends on the product required, the tree species and the growing conditions. It may range from less than four years to over thirty. A few trees (e.g. cherry and aspen) do not coppice properly but when the main stem is cut they sucker profusely from the roots. The suckering has been actively managed in some countries (Austad, 1988). Coppiced woods have a very clear and organised structure and appear to be very unnatural-looking but in terms of woody plant species they are sometimes considered to be closer to the wild-

··

Species Box 3.1: Hazel

Hazel is a shrub that usually has multiple arching stems and grows to 8 m tall. It is deciduous, with large leaves which are quite round in outline but with a distinctly pointed tip. The male flowers are spectacular yellow catkins that emerge very early in the spring and cast pollen. The female flowers are much smaller and have deep red stigmas. The fruits ripen in September and are ovoid nuts sitting in a large enveloping calyx. They are very palatable to grey squirrels that eat them before they are ripe, thus natural regeneration is limited on many sites. The nuts are edible and cultivated larger varieties have been developed. Hazel is widespread in Britain and is very hardy. It prefers fairly heavy but well-drained soil and is shade-tolerant. Coppiced hazel was a fundamental part of traditional woodland management and the rods were put to a very wide range of uses.

Source: White (1995)

··

wood than many more 'natural'-looking woods.

Coppicing dates back to Neolithic times and has been demonstrated to have taken place around 4000 BC. The system was well established in 1086 and by 1251 had spread to nearly all woods (Rackham, 1990). Most of the rotations were short at this time (4–8 years) but became progressively longer as the years passed. Coppicing on a large scale spread to the north rather later, the first cut in parts of Scotland was not until 1700 (mostly to supply the tanneries with bark and to make charcoal).

Coppice usually had some uncut trees or standards in each block which were a variety of ages. During the time of Henry VIII there was a legal requirement to have at least 30 standards per hectare. Today the canopy of the standards usually occupies 30 per cent to 50 per cent of the ground area, allowing plenty of light to reach the understorey.

The method of cutting hazel coppice used in Britain today is outlined below and is taken largely from Hampshire County Council's (1991) booklet. These suggestions are based at least partly on generating a commercial crop.

The hazel is usually cut between October and March as close to the ground as possible (other species are sometimes cut a little higher) as this encourages growth from the ground as well as from the stools. The cuts are angled outwards so that water does not collect in the stool. The suggested density of the stools is one every 8 m² or 1,250 per hectare. If they are not as dense as this, they can be encouraged by layering stems from existing stools. This is done by cutting part-way through a stem and bending it down along the ground. At the point where the new stool is required the stem is scored on the underside where it will touch the ground. Then it is pegged down using a piece of forked hazel. When the new plant is growing well the connection from the mother can be broken. The coupes should be no smaller than 0.5 hectare and no more than 25 per cent of the wood or 5 hectares.

The suggested number of standards per hectare is 50 and this is comprised of 20 saplings of 0–25 years, 12 young trees (25–50 years), 8 semi-mature trees (50–80 years), 6 mature trees (80–125 years) and 4 ready for felling (110 years plus). The standards should be oak or ash as these species are better than trees like beech (which casts a very dense shade on the coppice) or birch (low-value wood and which seeds rather too readily in the recently cut areas). It is useful to have species like crab apple, cherry, wild service, lime or hawthorn in the coupes if they are present in the wood and flowering shrubs and trees are important for some invertebrates.

Peterken (1992) points out that, traditionally, a coppiced wood did not just consist of the coppiced blocks, but that important elements were the grassy rides between the coupes and the ditches, hedges and banks bounding the wood and dividing it. In large coppices there might also be small meadows and perhaps ponds (see the case study of Bradfield Woods, p. 120).

In the early days of coppicing, the existing tree species were cut and there was largely no change to the species composition. By around 1800, many coppices were 'improved' by the planting of different species and became more formal in structure and lay-out (Peterken, 1996). The density of stools was sometimes increased (or dead ones replaced) by layering and planting, clearing was also carried out, removing unwanted species.

Coppicing declined from the late nineteenth century onwards. This was due to the drop in value of coppice products (coal and coke were widely available), the decline in need for hurdles, etc. (for the agricultural industry), and the difficulties of finding workers to cut the wood. This decline had two major consequences: first, the coppice itself was not cut for a long time. Many woods were last cut sometime between 1920 and 1950 (Peterken, 1992). Second, the management of the standards ceased so that those that exist today are much older than they would have been. If coppice was not simply neglected it was sometimes actively converted to high forest by increasing the density of oak (for the ship building industry) or encouragement of beech in the Chilterns for the furniture trade. The decline of coppice in Britain is illustrated by Figure 3.1.

There are several different types of coppice found today:

1 Simple coppice. An area with just coppice stools and where the coupes are all cut at the same age. It may have a mixture of species or may be just one (e.g. sweet chestnut).
2 Coppice with standards. Coppice as above but with a variable number of uncut standards or maiden trees. The standards were left for 3–8 coppice cycles (in the past up to 3 was more usual), ideally as a range of ages. The most characteristic species in this type are hazel as the coppice and oak as the standards.
3 Stored coppice. This became popular as

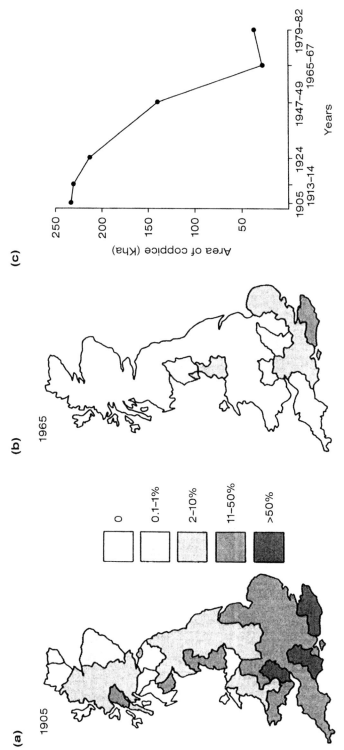

Figure 3.1: The decline of coppice in Great Britain. (a) and (b) The distribution of actively coppiced woodland in Britain in 1905 and 1965. (c) The decline in area between 1905 and 1979/1982. From Fuller and Warren (1993).

the working coppices declined. One stem of each coppice stool was selected and left while the other branches were removed. This process of 'singling' results eventually in a woodland resembling high forest but the bases of the trees are usually curved and sometimes uneven in shape (Plate 3.1).

4 Short rotation coppice. Defined as coppice cut on a rotation of less than ten years (usually shorter) for small produce. The products are now used for energy production and the tree species are poplar and willow.

Table 3.3 shows the amount of coppice found in Britain today broken down by country and species.

Plate 3.1: A 'singled' coppice, Buckinghamshire

Table 3.3: The area (hectares) of worked coppice in England by type and species and that in Wales and Scotland by type

	Principal species of coppice in England							Coppice in Wales	Coppice in Scotland
	Sycamore	Ash	Sweet chestnut	Hornbeam	Hazel	Other species	Total		
With standards	115	193	5 275	1 697	1 465	2 728	11 473	80	15
Coppice only	2 297	1 184	13 816	1 716	1 573	5 125	25 711	1 849	4
Total	2 412	1 377	19 091	3 413	3 038	7 853	37 184		
% of coppice total	7	4	51	9	8	21	100		
Stored coppice*							48 100	17 900	5 900

Source: Evans (1992)

Note:
* Estimate

Commercial coppice methods: Commercial coppice for fuel

There are two main reasons for coppicing today. The first is for purely commercial reasons. The second is done largely for nature conservation reasons or for both conservation and commercial reasons.

Short rotation coppice has been trialled and used experimentally as a method of producing wood chips for generating heat and power with willow or poplar. At the moment it is still in the developmental stage although some projects have been set up. One heat and power plant in Yorkshire aims to use 2,000 hectares of willow coppice to produce enough energy to supply 18,000 homes (Macpherson, 1998).

The trees are planted at high densities (12,000 to 15,000 per hectare), are coppiced at the end of the first growing season and then harvested mechanically every three years (Hart, 1991). The yield is expected to be about ten to twelve tonnes of dry wood per hectare per year if harvested every three to four years (equivalent to six tonnes of coal or four tonnes of oil (Hart, 1991)). Each crop is expected to stay in for thirty years before being replaced. The disadvantages are that irrigation may be necessary especially in periods of drought, and repeated removal of the vegetative material may lead to a reduction in nutrient levels over a period of time so fertilising is often deemed necessary.

Commercial sweet chestnut coppice

Sweet chestnut is the most important coppice crop in Britain but is mostly found in the southeast where the climate is warm and the sunshine hours long. It is grown for hop poles and fencing (chestnut paling). Rotations are between 10 and 20 years and the poles are harvested at a diameter of 7–10 cm. Most of the coppices were established in the mid-nineteenth century and so have been cut since that time (Rollinson and Evans, 1987). At each cutting, approximately one stool in 20 dies so the gaps are either planted up or a branch from a neighbouring stool is layered to fill the space.

Cutting hazel coppice commercially

There is a high quality market for hazel coppice and at present the demand exceeds the supply (Thurkettle, 1997). The produce is still used for thatching spars, charcoal, brooms, walking sticks, etc. and in 1990 there were 70 full and part-time underwood workers in Hampshire and the Isle of Wight so it is still an active industry.

If hazel is worked on a 6–10 year cycle, shoots of 4–5 m long result which are ideal for traditional crafts. About 25 tonnes dry weight per hectare can be harvested at 10 years (Hart, 1991). The Wessex Coppice Group has led the way in promoting and marketing coppice products. They consider that 4 acres (1.6 hectares) of coppice is needed for the underwood workers to make a living. One acre (0.4 hectare) of 8-year-old hazel coppice will produce 10,000–12,000 hazel rods 10–15 foot (3–4.6 m) in length which can be made into 300 six foot (1.8m) hurdles. In addition, there are also assorted bean and pea poles, clothesline props, etc. In Hampshire now there are about 271 hectares of hazel coppice in a worked cycle. Note that measurements of coppice produce are traditionally given in imperial sizes.

The effects of coppicing

The microclimate in coppice plots

An area that is coppiced suffers a dramatic change in environment, which has been likened to a natural disturbance by Evans and

Barkham (1992). The removal of the shrub layer and thinning of the canopy greatly increase the amount of light reaching the woodland floor. When there is little canopy, almost all the summer rainfall will reach the ground. However, due to the higher temperatures during the summer, increased light levels and greater wind speed, the rain will largely evaporate, thus the humidity levels will generally be low. Over time, as the growth on the coppice stools develops, the intensity of light at ground level will reduce, as will the amount of summer rainfall and the wind speed, hence the humidity levels within the woodland will rise.

Coppice and plants

Opening up the woodland may be detrimental to some species but it can also be beneficial to a wide range of woodland ground flora. Such 'coppicing' plants (Rackham, 1988) include primrose, bluebells and violets. They flower in profusion after coppicing, often giving the most dramatic result in the second or third year after cutting, and survive during the years of shading. Table 3.4 illustrates species favouring coppicing and those preferring non-intervention areas.

One problem in modern, recently cut coppices can be the tremendous growth of bramble (Species Box 3.2) in early years. While this will eventually decline as the shade from the coppice increases, it may be too late for the other ground flora. Bramble can be cut (and might well have been cut for faggots in the past, K. Kirby pers. com.) but this usually needs to be done manually to avoid damage to the coppice stools. It can also be sprayed but this is difficult because of the non-selectivity of the chemicals which kill bramble. K. Kirby (1992) has suggested that the increase in bramble and other nitrogen-loving plants may be due to the increased nutrients resulting from leaving dead branches on the ground rather than removing and using them as would have happened in the past. However, bramble responds strongly and positively to increased

Table 3.4: The plant species favoured by coppicing or non-intervention

Coppicing	Non-intervention	No preference
Common dog-violet	Dog's mercury	Bluebell
Early dog-violet	Great wood-rush	Wood anemone
Early purple orchid	Lily of the valley	
Giant bellflower	Ramsons	
Goldilocks buttercup	Sanicle	
Herb Paris	Yellow archangel	
Moschatel		
Pendulous sedge		
Pignut		
Primrose		
Purple small-reed		
Wild daffodil		
Wood speedwell		

Source: Evans and Barkham (1992)

Species Box 3.2: Bramble

Bramble is a well-known shrub, or undershrub, growing to 3 m in height. Many different forms can be distinguished, now considered microspecies, which are very difficult to separate so in most instances bramble is referred to as *Rubus section Glandulosus*. Brambles are very strong and prickly, can be dominant in woodlands and are found throughout the British Isles. The species is found on a variety of soils but tends to prefer those which are slightly acidic and avoids very wet areas. The stems may be perennial and the leaves can persist throughout the winter. Bramble flowers are white or pale pink and are produced between June and September. They are insect-pollinated but in some varieties fertilization is not necessary and the berries develop just from the female part of the flower. The berry is better known as the blackberry and, depending on the variety can have just a single druplet (seed surrounded by black flesh) or up to twenty druplets. Much of the seed is not viable so most of the regeneration is by vegetative methods. In the autumn the stems tend to grow downwards and when they reach the soil they produce roots

and a bud. Many of the fruits are eaten by birds, which disperse them, and at least some of these seeds do germinate (although this does not happen until two winters have passed). This being the main method by which bramble colonises new sites.

Source: Grime *et al.* (1988)

---

light levels. Martin and Martin (1995) showed that bramble increased from under 10 per cent cover to nearly 80 per cent between years three and four after cutting coppice in Lower Wetmoor Wood (Gloucestershire). However, by year eight the cover was back down to below 20 per cent.

Coppice and birds

Birds, being much more dependent on the structure of a wood, respond in a fairly clear way to coppicing and because they are more mobile they can seek out suitable areas easier than most animals. Figures 3.2a and b illustrates some of the species of birds found at different stages of the cycle. Note that this is simplified and that there are differences due to

regional and trees species factors. During years 3–8 the coppice is particularly important for several species of migrant. However, coppices are usually not very good for hole nesting species due to the lack of suitable old trees and rot holes. Standards are important for birds in providing feeding sites, potential nesting holes and perches but there needs to be a balance, too many standards and the coppice does not grow.

Coppice and mammals

A variety of species of mammal can be found in coppiced woodlands. Work done on small ground-running mammals at Bradfield Woods shows that the different aged coppice supports different numbers and species of mammal

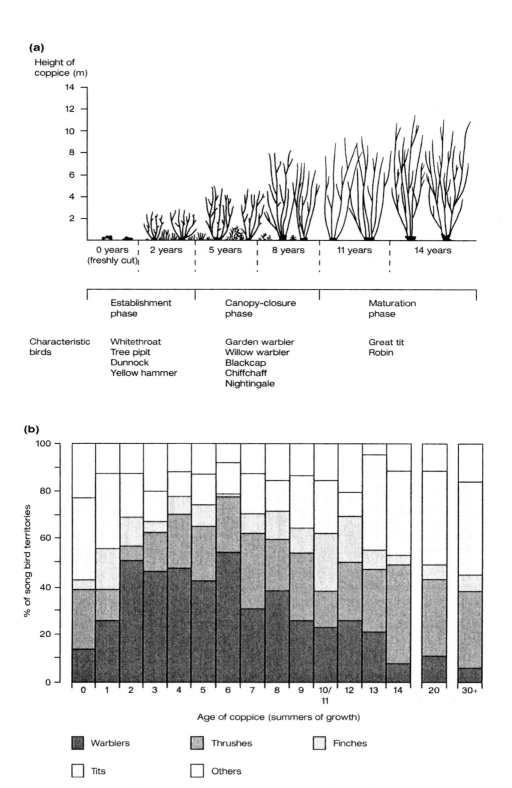

Figure 3.2: The birds found at different stages in a coppice cycle. a) An indication of the coppice structure and the most characteristic birds found. From Fuller (1995). b) Detailed information on the songbird communities in Bradfield woods in years following the cutting of coppice. From: Fuller (1992).

Plate 1: A Buckinghamshire beech wood in the autumn

Plate 2: Bluebells carpeting a woodland floor, Hertfordshire

Plate 3: Riverside woodland, Herefordshire

Plate 4: An old oak at Windsor Forest with *Ganoderma adspersum*, a heartwood decay fungus

Table 3.5: The relative abundance (% of total numbers) of different species of small mammal in different ages of mixed species coppice in Bradfield Woods (Suffolk) and an uncoppiced old oak woodland in Surrey

| Species | Age of coppice (years) | | | | | Oak |
	1	3	10	20	30	
Bank vole	10	37	19	16	10	41
Common shrew	23	25	17	17	14	13
Field vole	3	3	<1	–	–	<1
Harvest mouse	20	5	<1	–	<1	<1
Pygmy shrew*	<1	7	2	2	2	2
Water shrew	<1	2	<1	–	–	<1
Woodmouse	40	18	52	5	59	41
Yellow-necked mouse	2	2	10	11	14	4
Number of species >1%	6	8	4	5	5	5

Source: Gurnell *et al.* (1992)

Note:
* Probably an underestimate.

(Table 3.5). After cutting, woodmice are especially abundant but by three years after growth a wide range of species can be found with common shrews and bank voles the most abundant. In the old coppice where there is little ground flora the number of species declines (Gurnell *et al.*, 1992). Common dormice (Species Box 3.3) are another of the mammals most typical of coppice but they prefer older growth as they inhabit the branches of the bushes.

Generally, for small mammals Gurnell *et al.* (1992) recommend a mixed species composition with a mosaic of coupes of up to 30 years in age and consider that this type of management is beneficial to small mammals. Sweet chestnut coppice on its own is probably not particularly favourable.

Bigger mammals also use coppices and deer can have a very big impact. In the past, coppices were fenced, at least in the early years, to stop domestic stock getting in and this may have also deterred the deer. Now many are not

fenced and in addition deer numbers are probably higher. Deer like to have shelter from bad weather, cover to hide in and a range of different woodland successional stages to provide food, so coppices suit them very well. In fact, in other countries such as France and the USA coppicing has been done specifically to encourage deer (Ratcliffe, 1992). Roe deer and muntjac are the main species to exploit coppices and they selectively feed on vegetation that is high in nutrients and nitrogen (characteristics of the young leaves). Fallow deer can also have a big impact especially if there are open areas for them to graze either within the wood or close by. Coppice of 5–10 years in age supports the largest numbers of deer (Ratcliffe, 1992).

The management of coppice to reduce the impact of grazing by deer can take a variety of forms. Fencing is the most efficient but it is not always practical and may look visually obtrusive. Tree shelters are not effective against fallow deer and wire or polythene mesh guards

Species Box 3.3: Common dormouse

The common dormouse, sometimes called hazel dormouse, is a native British species. It is orange brown in colour and 60–90 mm long excluding the bushy tail. It should not be confused with the introduced fat, or edible dormouse, which is much larger and silvery grey in colour. Common dormice live in deciduous woodland where there is a dense scrub layer. The later stages of coppice cycles are ideal but they can also be found in a variety of other woodlands and in hedges. One of the most important features of the habitat is that it must have a wide range of food plants available during the summer months. The dormouse is well known to eat hazel nuts, which it opens by gnawing a round hole and leaving oblique tooth marks round the edge. It also feeds on other fruits and berries, flowers, pollen and buds in early summer and insects. One or two litters of about four young are born each year in a nest, about 150 mm in diameter constructed, up to 10 m above the ground, in dense undergrowth. The nests, mostly made of honeysuckle bark, may also include grass and moss. They are ball-shaped and do not have an obvious entrance. During the summer months dormice are entirely arboreal and nocturnal. The daylight hours are spent in nests (of which they may maintain three at any one time) and at night they climb amongst branches between 1 m and 12 m above ground. They are very agile. Dormice hibernate between October

and April, usually at ground level. Prior to hibernation they build up reserves of fat and may almost double in weight. This long period of inactivity has earnt them their reputation as sleepy animals but they probably wake up several times during the winter. Dormice can live for 4–6 years which is much longer than other mammals of a similar size. Now found from Leicestershire and Suffolk southwards the dormouse has declined in the last 200 years. Recent re-introductions are aimed at restoring the previous range.

Sources: Corbet and Harris (1991); Gurnell *et al.* (1992)

may work but need to be removed at a later date. Piling the brash on the cut stools stops browsing at the early stages of growth but often reduces the vigour of the growth by shading. Dead hedging round stools or coupes can work well but is very labour-intensive. (See the case study of Bradfield Woods, p. 120.)

Coppice and invertebrates

Perhaps the most obvious insects to be encouraged by coppicing are butterflies. A few species such as speckled wood and white admiral prefer shadier areas but a wider range, including many of the fritillaries (Species Box 3.4) benefit in the early stages of the coppice cycle. Nearly all the losses of butterfly species from woods have been attributed to the decline in coppicing. Figure 3.3 shows the numbers of butterflies found at different stages in the coppice cycle from Hampshire and illustrates their dependence on the very early stages.

Species Box 3.4: Pearl-bordered fritillary

The pearl-bordered fritillary is one of four species of woodland fritillary that feeds on violets.
All were quite common in the past but have suffered large declines since the 1950s. The pearl-
bordered fritillary is found in fairly open areas
and seems to particularly like places where the
trees have been cut or coppiced just a year or so
previously and violets are growing in profusion.
The adults are an attractive, speckled orange
and black with paler undersides including some
white patches and 'pearls' around the edge.
They emerge in late April or early May and the
males are usually quite conspicuous. The
female lays single eggs on or near plants of
almost any species of violet although young
common dog violets seem to be preferred. The
caterpillars hatch and start feeding in June and
then hibernate over winter after the fifth moult.
They emerge again in March to feed for another
month before pupating. Coppicing suits this butterfly very well because it prefers recently created clearings. Due to the decline in coppicing,
the pearl-bordered fritillary has a much-

restricted distribution now and is found mostly in
the south and west but also in Scotland. It seems
quite slow to re-colonise suitable sites.

Source: Thomas and Lewington (1991)

In general, the higher the density of standards, the fewer the butterflies, because the
standards shade out their food plants. As butterflies complete their life cycle in one year,
they need a regular and plentiful supply of
food plants and therefore ideally a new coupe
should be cut each year on a short rotation.
The coupes are best interconnected by wide
sunny rides to aid dispersal (Warren and
Thomas, 1992).

Moths feeding on plants that flourish after
coppicing are generally promoted in the early
stages but there are others feeding on trees and
shrubs that prefer the older stages of the cycle.
Table 3.6 illustrates the species differences
associated with recently coppiced areas and
neglected coppice.

The effect of coppicing on other
invertebrates is much less well known. The
bare ground created initially suits ground living groups such as ground beetles and wolf
spiders with the highest numbers being
recorded two and five years after cutting. The
ground running fauna and that of the low
vegetation were studied by Steel and Mills
(1988) by pitfall trapping and UNIVAC sampling at different stages of the coppice cycle
(see Figure 3.4). As can be seen, the number of
individuals peaked in the season following
cutting for the ground running fauna but a little later for the fauna of the low vegetation.
The number of species, however, was much
less variable, especially for the ground running
fauna.

The abundance of flowers and the warm
microclimate do suit some species very well.

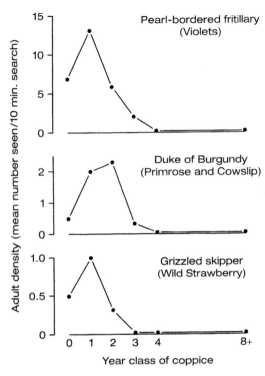

Figure 3.3: The density of three species of butterfly in hazel coppice of different ages. Their larval food plants are indicated in brackets. From Warren and Thomas (1992).

But many species are not favoured by coppicing because there is not enough old and decaying wood around to provide suitable habitats. The diversity of species found as the hazel coupes get older does not decrease as dramatically as other groups of organisms. This is because the structural diversity of the trees and shrubs increases (e.g. there is more dead wood on the ground and in the underwood) and some species of invertebrate can make use of this (Southwood *et al.*, 1979 cited in Greatorex-Davies and Marrs, 1992). Welch (1969) showed that while most insects depending on hazel could be maintained on a 10-year cycle, some needed 30–40 years (e.g. wood boring beetles). In addition, some

spiders have been found to be more abundant in older coppice because of the structure of the bases of the stools (Sterling and Hambler, 1988).

The young, nutrient-rich leaves that the deer find palatable are equally well liked by many herbivorous insects. Clearings in woods are especially good for some Hymenoptera, probably due to the sheltered conditions and high temperatures. Many species like the edges of woods and the rides so a good mosaic of coppice suits them.

Management for conservation – to coppice or not?

From the preceding sections it is apparent that coppicing will favour some organisms but not others. In addition, the many variables such as the species of tree, length of rotation, density of stools and standards mean that coppiced woods can vary in appearance.

Coppicing to suit dormice may involve a fairly long rotation time of 10–15 years or more, with narrow rides and paths between the coupes and plenty of overhanging branches linking the woodland on either side. In contrast, coppicing to suit butterflies would be best on a shorter rotation with wide, grassy rides between the coupes. If commercial production is also required, this might fit quite well with management for butterflies but less well with that for dormice.

One of the important points to be established about coppicing is why it is being carried out and what is it aiming to benefit. Once the interest of the site and the aims are established, it is easier to draw up a plan of how to manage the area. Obviously there are occasions when compromises have to be made but even if both dormice and early rotational butterflies are present, answers can sometimes be found. For example, at

Table 3.6: Responses of moth species to coppicing compared to overgrown woodland

Number of individuals significantly greater in recent coppice in both years sampled	Number of individuals significantly greater in recent coppice in one of the two years sampled	No significant difference between numbers in coppice and overgrown woodland in either year	Significantly fewer individuals in the recent coppice in one year	Significantly fewer individuals in the recent coppice in both years
Brown Rustic	Heart and Dart	Chocolate Tip	Brindled Beauty	Dotted Border
Great Prominent	Oak Lutestring	Common Swift	Clouded Drab	The Dun-bar
Small Dotted Buff		Dark Arches	Common Quaker	Green-Brindled Crescent
		Figure of Eight	Copper Underwing	July High Flier
		Frosted Green	The Engrailed	Marbled Minor
		Maiden's Blush	Feathered Thorn	Mottled Beauty
		November Moth	Hebrew Character	Small Quaker
		Scarce Umber	Ingrailed Clay	Scallop Shell
		Small Brindled Beauty	Large Yellow Underwing	
		Spring Usher	Mottled Umber	
		The Chestnut	Pale Oak Beauty	
		Willow Beauty	Popular Hawk	
		Winter Moth	Riband Wave	
			Small Fanfooted Wave	
			The Sprawler	

Source: Waring (1988)

Note: The species shown are those which were trapped in sufficient numbers for a chi squared test to be calculated to compare the numbers found in recently coppiced woodland with those from overgrown broadleaved woodland in Bernwood. All levels of significance taken as 0.05.

Hatfield Forest and Bradfield Woods, two coppice rotations are worked, a short one and a longer one.

There has been much recent discussion in the debate for and against coppicing. Hambler and Speight (1995a and 1995b) argue against traditional management techniques carried out for nature conservation reasons on the grounds that true woodland species are unable to survive the dramatic clearance effect and only robust species benefit. The counter-argument, that many typical coppice species are those requiring clearings which are lacking in many woods, and that some species are now really only found in actively coppiced woods is

put by Bignal *et al.* (1995) and Kirby (1995). Kirby (1984), Peterken (1996) and Fuller and Warren (1993) also cover this subject in more detail.

Figure 3.5 illustrates the type of thought processes a site manager should go through to decide whether to coppice a woodland or not. The weakest link is usually the survey of what lives there. This is because to compile a reasonable species list requires the work of specialists and so can be expensive. Thus 'easy' and popular groups (e.g. flowering plants and butterflies) tend to be better represented and the management is more likely to be carried out for them than for saproxylic beetles or

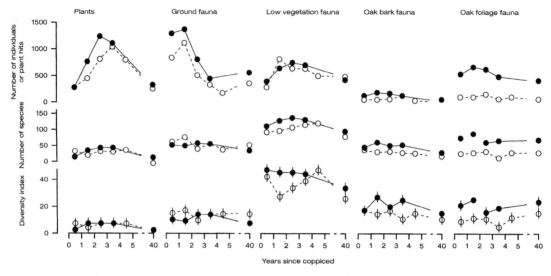

Figure 3.4: The diversity of plants and invertebrates in coppice of different ages at Brasenose Wood. Open circles represent 1980, filled circles 1981. The diversity index used is that of Williams ± standard deviation. From Steel and Mills (1988).

flies. Historical records can be useful but need to be used with care. The presence of coppice-favouring species many years ago is not necessarily a justification for coppicing again after a long lapse. If those species are now extinct and deep woodland species have spread from adjacent woods, the important species today might be quite different from those recorded in the past. Of course traditional management is sometimes justified in its own right for historical and cultural reasons, e.g. Bradfield Woods (Kirby, 1988a).

If the main purpose of the management is for commercial reasons, then the thought processes may be slightly different. If there is a market for the coppice produce, the continuation of an existing rotation may be possible. If the market is only for timber, the coppice can perhaps be stored. Ideally the objectives in terms of nature conservation should be identified first and then adapted as little as possible in the light of commercial interests but this is not always feasible.

WOOD-PASTURE, POLLARDS AND DEAD WOOD

What is wood-pasture?

In its simplest form, wood-pasture is woodland that is grazed. The grazing land under the trees was grassland or heathland and the practice dates back to Anglo-Saxon England as *silva pastilis* (Rackham, 1988) usually on wooded commons. The local people had rights to graze their animals on the commons and sometimes also to cut wood. In order to combine grazing and wood production, one of two practices was usually followed. Either the common was compartmentalised so that the stock was kept out of recent coppiced areas, or the trees were pollarded. It is the latter situation we are mostly concerned with here. A pollarded tree is one which has had the branches lopped off at a height which the animals grazed underneath cannot reach. Like coppiced trees, a crop of branches grows from

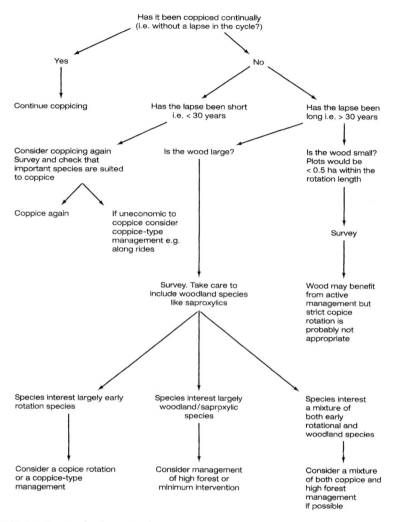

Figure 3.5: How to manage coppiced woodland.

around the point of cutting, but unlike coppice this cannot be browsed because the stock (or deer) cannot reach it. Young trees were probably pollarded at an age of 25–35 years old and subsequent cuts made at intervals of 12–15 (Le Sueur, 1931). In many places it seems that the cutting of the trees was not carried out in regular blocks like the coppice coupes and at least in some tree species not all the branches were removed at any one time. The trunk of the tree, or bolling, lived for many years and the branches were used for a variety of purposes depending partly on the tree species. Examples of use being domestic fuel, fuel for bread ovens, small-scale building, etc. In other countries (e.g. Sweden and Norway) the branches were cut during the summer and the leaves dried like hay and fed to the animals as winter fodder (Austad, 1988). This may have happened to some extent in Britain (see Rackham, 1988) and certainly pollarding for fodder took place in the Lake District and has recently been reinstated in the New Forest where holly is cut to feed the ponies in the winter. Pollarding for the primary purpose of producing animal fodder was probably not as widespread in Britain because the climate allowed hay to be more readily available.

In medieval times some of the wooded commons were incorporated into the parks and Royal Forests and were subsequently grazed by deer. More recently, in the eighteenth century, some of these were landscaped into formal settings by Capability Brown and Humphrey Repton who often made use of existing older trees in their designs. Some of the wood pastures are still grazed today, notably parts of the New Forest, also Moccas (Herefordshire) and Dunham Massey (near Manchester) but many are no longer grazed.

A rather different type of woodland grazing does take place today especially in the uplands. These woods are usually part of a farm and are grazed because of the need to increase the area of land for commercial stock. Many of these woods are being grazed much more heavily now than in the past but this will be discussed in more detail in a later section.

Old trees

A variety of different species were pollarded in the past, including oak, ash, hornbeam, sweet chestnut, willow, beech and even crab apple and hawthorn. One interesting feature of pollarding is that trees that have been regularly cut live considerably longer than uncut trees. This is also true of coppicing but old coppice stools tend to be less ecologically valuable than pollards because of the smaller amount of dead wood. For example, a maiden beech tree might live for 250 years whereas beech pollards at Burnham Beeches can be up to 420

years old (Read *et al.*, 1991). Not all old trees are pollards. In a wildwood situation some trees would have survived to a great age naturally.

There are no easy definitions of an ancient, veteran or old tree, partly because, dependent upon the species, the characteristics of old age occur at different times. Thus a birch at 80 years shows the rot and decay expected of an old tree whereas an oak at 100 is barely mature. Oak trees in particular can live for many years as old trees. They are able to 'self-pollard' by shedding branches naturally so that the trees may become wide, squat and hollow but remain quite healthy. Perhaps the best definition of a veteran tree is that it is of interest biologically, culturally or aesthetically because of its age or size.

Because of the great age of the trees they are likely to be the best genetic link we have to the trees of the wildwood. An older tree develops various other characteristics, which are closely correlated with its value in terms of nature conservation:

1 The heartwood starts to decay and rot (Plate 4). This rot begins with the fungi (most of which are not pathogenic but just decay the wood, for example chicken of the woods, Species Box 3.5) and the rotting conditions are suitable for a wide range of invertebrates to live in, some of which also contribute to the rotting processes. The tree may feed itself from the rotting heart wood by growing aerial roots down from the branches.

2 As the rot becomes more extensive, large holes develop which are suitable for hole nesting birds and as bat roosts.

3 Cavities and hollows fill with water, providing a different range of habitats for insects with aquatic larvae.

4 The bark of the tree becomes loose in places

Species Box 3.5: Chicken of the woods

Chicken of the woods (also called the sulphur polypore) is a large (0.1–0.4 m) bracket fungus. It is irregular in shape, though consisting of roughly semi-circular layers, with a lumpy texture. The most outstanding feature is the bright egg-yolk yellow colour. When fresh, it is soft and fleshy but dries to a whiter, more crumbly structure as it ages. The spores are produced from tubular pores and are white in colour. The chicken of the woods is found on a range of deciduous trees but is perhaps most frequent on oak. It is a heart rot species that lives in the tree, causing it no harm, but breaks down the dead wood. The fruiting bodies do not indicate that the tree is likely to die.

Source: Phillips (1981)

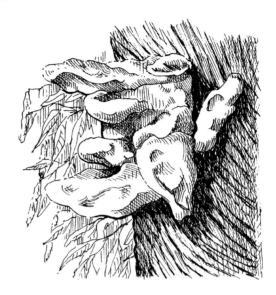

creating suitable niches for invertebrates and bats.

5 Die-back in the crown results in dead branches, homes for more invertebrates and vertebrates.

6 Sap runs down the outside of the trunk to provide feeding sites for invertebrates and moist regions for mosses.

7 The range of surfaces suitable for epiphytic bryophytes and lichens increases.

8 When branches break off they leave shattered ends which provide suitable surfaces for fungi and invertebrates to colonise (better than cut surfaces).

If the old tree is still alive it will continue to provide these features into the future. If it dies, it is still valuable while it slowly decays. Thus it is best left in place and not removed. Likewise, dead wood on the ground can provide suitable habitats for some specialist fungi and invertebrates. So the provision of dead wood is important especially in sites where the fauna associated with old trees is present.

Organisms associated with dead wood are sometimes called saproxylic species (Plate 5). This term covers a wide range of 'life styles' including the wood rotters and wood-decaying species, those found on the wood-rotting fungi (for example some of the fungus gnats, Species Box 3.6), and even carnivores which are restricted to old trees and dead wood. Some species are relatively generalist, for example the millipede *Cylindroiulus punctatus* (Species Box 3.7) which is found in almost all rotting wood of a comparatively young age, both on trees and on the ground. Others are incredibly specialist, being restricted to specific types of rot holes on specific tree species.

Threats to old trees and woods of old trees

Britain has a relatively high population of old trees in comparison to other European

- -

Species Box 3.6: Fungus gnats

The family of true flies known as the fungus gnats includes about 500 species, many of which have larvae that eat fungi. The adult flies are quite small and delicate, quite gnat-like in appearance and are often humped in shape. The larvae are mostly white with darker heads and may be restricted to specific species of fungi or specific types, such as those growing on wood. Within a fungal fruiting body a succession of different species of flies may be found (not just the fungus gnats) as the fungus grows, ripens and then decays.

Sources: Chinery (1976); Stubbs and Chandler (1978)

- -

countries. This is largely due to our social history, which has perpetuated them by pollarding, and protected them from removal by including them in parks and on wooded commons. The lack of wars being fought on British soil is also a major contributory factor to the retention of old trees relative to much of the rest of Europe. However, old trees are not safe from a variety of threats. These threats include felling, conifer planting, natural regeneration and inappropriate management of surrounding land.

Felling

Large numbers of old trees have been felled for many different reasons. In the past they were cut down as woods were cleared to make room for development or agricultural intensification. Although this undoubtedly still happens, probably a bigger threat today is felling for health and safety reasons. Increasing awareness of owner's liability together with increased tendency to sue for compensation have made the owners of many old trees nervous about them. The risks attached to old

Species Box 3.7: *Cylindroiulus punctatus* and other millipedes

Millipedes are detritivores, feeding largely on dead and decaying plant matter, breaking it into smaller fragments. Several species are commonly found in woodlands feeding either on the fallen leaves or on dead wood. *Cylindroiulus punctatus* is a very common species, found in almost every piece of rotting wood. It feeds between the bark and the wood proper and can be easily discovered by folding back loose bark on decaying logs. It is a snake millipede (or julid) and, as the name implies is cylindrical in shape. When disturbed it curls into a characteristic spiral shape. This feature together with its size (14–27 mm), its pale, or sometimes banded, appearance and the blunt stubby tail (seen when viewed with a hand-lens) make it easy to recognise. Millipedes start life as eggs, laid in a small cluster. These hatch into a short-lived, immobile 'pupoid' and then into a tiny version of the adult with just 6 legs. At each of their numerous moults, snake millipedes add more body

segments (and hence legs) and also eyes. *Cylindroiulus punctatus* takes three years to reach maturity. Females may then survive and breed in subsequent years but males only breed for one year before dying.

Source: Blower (1985)

trees in the middle of woods, on private property, are usually very low. The chances of branches or trees hitting people, cars or houses are increased should the tree be next to a public footpath, car park, neighbouring property or in a well-visited park or common.

In many situations, felling the trees can be avoided, for example:

1 the element of risk may be reduced by altering the route of the footpath or relocating the car park (as has been done in the New Forest);
2 carrying out tree surgery but doing as little as possible to make the tree safe (e.g. removing overhanging branches or cutting the tree back rather than felling it);
3 discouraging people from going too close to the tree, e.g. fencing particular notable trees or using more inventive methods. At

Dunham Massey a 'deer sanctuary' is marked out and people are requested to refrain from crossing a low, simple barrier to enable the deer to obtain some solitude away from the large number of visitors. This area has been chosen because it is where the majority of the old trees are and this lessens the potential risk of branches falling on visitors.

Planting up areas of old trees with conifer plantations

Commercial forestry in places such as Sherwood and Windsor Forests has resulted in the extensive planting of conifers in between the old trees (especially oak). Many old trees have survived but are now being shaded out by the conifers. On both of these sites the conifers are now being removed from round the old trees to help them survive. It has been found that sudden removal of the planted trees can sometimes cause desiccation of the old ones. If possible it is probably better to thin the conifers over a period of years rather than fell them all at one time.

Natural regeneration competing with the old trees

Grazing has ceased at the majority of sites which were formerly wood-pastures resulting in severe competition from young secondary woodland. The older trees suffer especially if their crowns are reducing naturally (or being cut). This naturally regenerated woodland can be removed but, if it is well established and very dense, it may be best to thin it over a period of years.

Crown imbalance

When trees were pollarded regularly the branches would not have become particularly large. Since the cessation of regular pollarding, many branches on pollards are now very large and heavy (up to 200 years old) and threaten to tear the tree apart. With oak this appears to be less of a problem because the trees shed branches naturally and recover. Other species may need careful tree surgery to relieve the weight in the crown. This does not mean removing all the branches as this usually kills old trees. If the tree is an appropriate shape, selective thinning of the branches is usually successful. It is best not to cut right back to the bolling (or trunk) but leave a stub to encourage regeneration. As a general rule, work should not be done in the autumn or spring but preferably between Christmas and March. Wherever possible, periods of heavy frost, previous drought conditions (or a predicted dry summer) should be avoided.

Succession

In wood-pasture that is actively grazed there are rarely many younger trees of the same species as the older ones because the seedlings are eaten by the herbivores (Plate 3.2). This can

Plate 3.2: Old oak trees at Staverton Park, Suffolk

also be a problem where plantations have been planted round the old trees and where natural regeneration of secondary woodland has occurred. Thought needs to be given to the next generation of old trees. For most saproxylic species and many epiphytic lichens and bryophytes, the continuity of habitat is very important. Most are not very good at dispersing so they require suitable conditions very close by. As the old trees decline in numbers, there is often a lack of suitable habitat for many of these specialist species.

To solve this problem grazing may have to be reduced or excluded from certain areas. Trees can be planted (and protected if necessary) using local stock if possible. Saplings can also be moved from elsewhere on the site. Commercial 'tree spades' can now dig up trees with a large amount of root and soil which increases their chances of survival and gives them a better start. This has been done successfully at Ashtead Common (Surrey). Where possible, the tree species should be planted in similar proportions to the old trees. Even when younger trees of the same species are present, they may take many years to develop the characteristics of the veteran trees. A solution to this problem has been attempted by pollarding young trees, and deliberately damaging trees, to introduce some of the rotting processes. This may obviously conflict with any requirement to produce good quality timber trees. New pollards are best located in the proximity of old ones.

Inappropriate management of the land surrounding the trees

There may be threats to the wildlife interest of old trees through bad grazing management. These include: animals stripping bark off the trees; trampling and increased nitrification of the area round the bases of the trees; and the

use of chemicals (in particular ivermectins) for worming domestic animals which may have a detrimental effect on free-living invertebrate populations as well as internal parasites.

Some wood-pasture sites where grazing has ceased have an almost continuous cover of bracken between the trees (e.g. Staverton Park, Suffolk and Ashtead Common, Surrey). The bracken is a problem because the dry leaves and stems on the ground in the late spring are a fire hazard. At Ashtead Common, fires have killed over 100 old oak pollards in recent years (Alexander *et al.*, 1996). Firebreaks have now been created in the bracken by spraying with Asulox, cutting and rolling.

Cultivation, such as ploughing, damages the tree roots and the use of fertilisers, herbicides and insecticides has detrimental effects on the organisms associated with old trees. Compaction by people, animals or machinery is also damaging.

Management of dead wood

Dead wood on the woodland floor (Plate 3.3) can make an important contribution to potential habitats for saproxylic species. Although some dead wood is more valuable than others, the availability of mixed ages and species is always valuable. The following principles apply even in woods without a specialist dead wood interest.

1 Bigger is better, both in diameter and length. The larger the piece of wood, the more valuable it is. Smaller pieces have a bigger surface area to volume ratio and dry out more quickly. Most invertebrates prefer damper wood.

2 It is best to leave the piece of wood where it falls. If it has to be moved, take it as short a distance as possible. The best location is a

Plate 3.3: Dead wood on the woodland floor, Suffolk

semi-shaded one, but a fully shaded place is usually preferable to somewhere in full sunlight.

3 If there are large quantities of logs, they may be piled up, again best in partial shade.

4 Large quantities of brash, smaller twigs and branches can be made into habitat piles. They are of greater value to birds and mammals than invertebrates. A small number of big piles are better than many small ones. They are more valuable if the brash is tied together in tight bundles before stacking because the moisture levels remain higher in the bundles.

5 When freshly cut timber has to be removed from the site it should be done as soon as possible after felling. If cut in the winter months, it should be removed before the end of April or left permanently. By May the invertebrates are active and potentially laying eggs in the log. Wood removed after this destroys part of the population of invertebrates. Felled timber with evidence of rot should always be left on site and not

destroyed. If timber has to be left to season on site before sale, it should be stacked and covered, covering prevents the entry of invertebrates (and speeds up the seasoning process).

6 Burnt and wet wood should be left if possible. Do not start a new fire with the charred wood from a previous one as some invertebrates live in burnt wood.

7 Dead standing trees should be left wherever possible.

8 Stumps should not be treated. If they are jagged or uneven they should be left and not be tidied up after felling/falling.

9 Consider deliberate 'damage' to trees to make them better for dead wood invertebrates, e.g. drilling holes, breaking branches and ring barking unwanted trees rather than felling.

10 Nectar sources. While not strictly the management of dead wood, many saproxylic invertebrates have a larval stage feeding on dead wood and adults which are nectar feeders requiring flowering shrubs.

11 Fungal fruiting bodies should not be removed nor trees felled because they have bracket fungi on them. The fungi are usually just rotting the heart wood and do not necessarily indicate that the tree will die. Many invertebrates live in fungi.

12 Dead branches on live trees should not be removed unless necessary for safety reasons.

P. Kirby (1992) and Fry and Lonsdale (1991) give more details about the management of dead wood. Alexander *et al.* (1998) give detailed guidelines on the management of ancient trees and dead wood.

Old trees and wooded landscapes in parkland

Much of the preceding discussion about the management of old trees and wood-pasture applies to parkland situations, but here there may be an additional consideration. Some of Britain's foremost wood-pasture sites were, or are still, part of the formally designed landscapes attached to large stately homes. Because of this there may be strong arguments for clearing away all dead wood on the ground, felling old trees to plant new ones or to remove them from parts of the landscape designed not to have trees. As a compromise, it is often possible to keep areas close to the house 'tidy' and further away standing dead and dying trees can be left together with wood on the ground. One of the biggest problems in parkland is the lack of the next generation of trees and here planting is often beneficial to both conservationists (to provide a continuity of habitat) and landscapers (to restore particular features, views or avenues which have been lost because of tree death/felling).

MANAGEMENT OF HIGH FOREST

High forest is generally considered to be a stand of trees grown from seedlings rather than from coppice shoots (Peterken, 1993). In general, the active management of high forest is for commercial reasons, i.e. to obtain a timber crop, and the extensive area of Britain under this form of management is outlined in Chapter 1. However, not all high forest is currently being worked commercially and much, especially that composed of broadleaves, is just neglected. While much high forest has been planted, other woodlands have developed naturally to form high forest. Sometimes particular tree species have been selected as the woodland develops. For example, most of the Chiltern beech woods gained their character as the beech was selectively encouraged for the local furniture trade (Countryside Commission, 1992). In addition, some high forest is young secondary woodland.

Secondary woodland

There is much woodland in Britain that is relatively young, having developed on open ground, and consisting primarily of pioneer species of tree. While often this type of woodland is not managed, largely because neglect is often the reason that it has developed, sometimes management is resumed. There are three broad methods of managing secondary woodland. It can be left to develop as a 'natural' succession; it can be managed in order to select a particular species or combination of species (either the pioneers or a more profitable timber crop); or it can be felled to restore the open habitat. The latter is often preferable if the situation where the trees are growing is of higher conservation value than the woodland developing (for example, heathland or moorland). As the young woodland develops from open ground it passes through a scrub phase. The management issues concerned with scrub will be discussed in *Grassland Habitats* in this series.

Commercial forestry

Commercial forestry covers substantial areas of Britain and the management of these forests has a significant effect on the environment within them. It is important not to neglect this form of management when thinking about nature conservation.

At its simplest, commercial forestry involves planting trees, carrying out a small amount of management as they grow and then harvesting them. This system will be looked at first, and some variations covered afterwards. Note that the term forest used in this context does not mean the Royal Forests as outlined in Chapter 1.

Planting

This is still the most common method of restocking woodland and many commercial forests are, or have been, planted on areas that have been open for many years. This afforestation is particularly pronounced in upland areas where the land is very poor and not suitable for other uses, such as agriculture. Trees are usually obtained from nurseries at 2–3 years old and are planted in early or late winter when they are dormant and the ground is not frozen. They are usually planted 1.5–2.0 m apart (4,400–2,500 per hectare, Kirby, 1984) in rows as this makes further operations easier. The trees are either all the same species or a mixture, with mixtures usually planted in blocks or rows. When planted, the trees usu-

ally need to be protected from browsing by deer and rabbits. Other plants may need to be discouraged by screefing (removal of the top layers of vegetation to the mineral soil), scarifying (similar to harrowing) or ploughing before planting and controlled afterwards by chemicals or physical removal. Fertiliser is sometimes added to poor soils in the form of phosphate. The use of Tuley tubes for broad-leaved trees helps to protect them, encourages growth by providing warm, humid conditions and helps to identify young trees amongst the weeds. Trees usually grow faster in the first three years if grown in tubes.

Sometimes trees are planted with a nurse crop. These are faster-growing species, which may help the trees they are planted with in up to four ways, by:

1 suppressing competing vegetation;
2 modifying the microclimate to help protect tender species from frost and exposure;
3 improving the soil nutrition for the other species, for example, European larch is used as a nurse crop for beech;
4 encouraging the main crop to compete for light, thus producing clean straight stems.

Nature conservation implications of planting

Planting trees on a previously open area will substantially change its character. This should be avoided if the area has a high conservation value (e.g. lowland heath or the Scottish Flow Country). Planting on land with a low conservation or agricultural value may be the best use for it, especially if the planting is done sympathetically. Alterations to the aesthetics of planting blocks by softening the shapes and mixing the composition generally improve the conservation value of commercial plantations, for example, sinuous boundaries are visually better than straight edges and this increases the length for edge species (Smart and Andrews, 1985).

Ground preparation for tree planting can be detrimental, especially if carried out in ancient woodlands with the aim of converting it to a commercial plantation. Many of the animals and plants living in the wood will be less disturbed by the encouragement of natural regeneration and selective felling rather than blanket planting. Weeding and fertilising can have effects on the invertebrates and birds, etc. as well as the plants.

The trees planted in forestry operations are often grown from imported seed but increasingly nurseries will grow plants from local seed sources and these are always preferable from a conservation point of view. Imported stock may sometimes be better from the timber quality point but local plants are likely to be better adapted to the local climate and the use of them helps to maintain genetic variety.

Thinning

As the trees grow they pass through several recognisable stages which can be identified as establishment, pre-thicket (1–3 m tall), thicket (to 10 m when the canopy starts to close over), pole stage (10 m to maturity), mature (close to felling age) and retention (kept beyond normal rotation). The animals and plants associated with these stages will change as the density of the canopy changes.

Conifer plantations show a succession of bird species occurring from meadow pipit and woodlark in the early stages to whinchat and willow warbler at around 10 years, followed by chaffinch, coal tit and crossbill when in the maturing phase. Probably the more varied the field and shrub layers in the early stages, the higher the density of birds (Fuller, 1995). Birds

later in the cycle are canopy dwellers, feeding on tree seeds or insects and the two most widespread birds of prey are usually tawny owl and sparrowhawk. The dominant species of tree makes a difference to the densities of birds, spruce generally being better than pine. Common crossbill and siskin have substantially expanded their ranges in recent years as a direct consequence of the increase in maturing conifer plantations.

The ground flora also changes as the trees grow (Peterken, 1996), though the species composition will depend on the site. Foxglove seeds can survive in the soil through a long rotation and grow when a gap is formed or when trees are felled. Also in clear fells rosebay willowherb and bracken may grow well. After about 20 years the vascular plants virtually vanish but the bryophytes increase during the dark closed canopy stage. Some plants do survive under the dense canopy, e.g. Broad buckler-fern, bramble and wood sorrel (Species Box 3.8).

As the trees grow they are usually thinned at various points to enable a good final crop to be produced (Plate 3.4). This also provides an income part way through the rotation. The density of the trees is reduced to about 70–100 trees per hectare for broadleaves and 200–400 for conifers. Thinning either removes selected (usually the poorest) trees or complete lines creating more space for the remaining trees. As well as thinning, other silvicultural operations are usually carried out. Brashing removes the dead lower branches of coniferous trees and allows easier access for thinning, etc. Pruning removes live branches and improves the timber value of the crop by reducing the number of knots and forks. It is usually done only on high quality value crops. Cleaning is the removal of other woody species, which are not required, e.g. self-sown trees and regrowth from cut stumps.

Species Box 3.8: Wood sorrel

The dainty wood sorrel is widespread in Britain and found in a range of moist woodland situations often where conditions are acidic. It is characterised by its three lobed bright green leaves, which are like those of clover except that they droop. The flowers are white (sometimes with pink tinges), simple and arise singly on long stalks in May. They are pollinated by insects but only rarely, so that seed is not usually produced. Later flowers are also produced, these have no petals, but are fertile. The plants form a rosette and then spread out, by rhizomes, to form large patches. Wood sorrel is very shade-tolerant and the leaves persist for a long period of the year. The leaves contain oxalic acid, which is poisonous to grazing animals.

Source: Grime *et al.* (1988)

Nature conservation implications of thinning

As a consequence of thinning, more light reaches the floor of the wood and this encourages a greater diversity of ground flora. Thinning tends to remove the poorest trees, which are generally those good for invertebrates. However, the stumps and piles of brash left behind may be beneficial.

If one species is selectively removed (as in the case of a nurse crop), the diversity of the trees may be reduced, but often the faster-growing trees are of less nature conservation value than the slower-growing ones, which are retained. In general, heavier and early thinning is best for nature conservation.

Harvesting

Harvesting is the felling and removal of the main timber crop. Felling may be done by hand, using chain saws, or using a harvesting machine (Plate 3.5). If felling is carried out close to an access route, the timber can be cut to length and stacked at the roadside by the felling machine. In situations where it needs to be extracted from the middle of the forest, a fowarding machine is usually used but, in inaccessible areas or on sensitive sites, horses are a useful alternative for small-scale removal (Plate 6). Felling is normally carried out when the growth of the trees has reached their maximum mean annual increment and thus the crop is at its commercial peak. The age of the trees can vary widely: 40 years is the quickest for conifers (normally 50–60); with deciduous trees it may be 120–150 years (Peterken, 1993). Thus the trees are cut long before they would have died of old age. Harvesting may open up large areas of the forest very quickly and has a considerable disturbance effect on the wildlife. Harvesting timber trees may have a similar effect to coppicing but there are clear differences: the rotation length is much longer than most actively worked coppice; the sites are not usually ancient woodland; the area

Plate 3.4: A thinned plantation of Scots pine near Aviemore

Plate 3.5: Harvesting timber by machine at Kielder, Northumberland

cut is usually much bigger; and often the trees include conifers. Thus the canopy is very dense before felling and the opening up effect more extreme. The consequences can include severe desiccation and exposure and the plants colonising these areas tend to be opportunists and good colonisers rather than woodland species. There are several different types of harvesting techniques, which result in different types of forest structure.

Even-aged high forest

The trees are usually planted at the same time and grown to maturity. They are then all felled at the same time and after felling the area is replanted with another crop of trees. Although this is usually the least desirable system for many organisms, because of the extreme change in microclimate, there are many pioneer species that can make use of recently cleared areas. In addition, some species of nature conservation interest like cleared areas that are subsequently planted. A notable example is the woodlark (Species Box 3.9). This bird colonises clear fells very quickly but rarely uses them beyond five years after replanting (see Figure 3.6). Nightingales also use the early stages of plantations but prefer areas with up to 3–5 years of growth and can tolerate older growth.

Species Box 3.9: Woodlark

The woodlark is brown in colour, looks rather similar to the skylark but is smaller (150 mm long) and has conspicuous white eye stripes that meet at the back of the head. Its song is less varied and powerful than the skylark. While singing it soars in wide spirals and then plunges down towards the ground with closed wings. There are two main populations of woodlark in Britain, in the south and east where it uses young coppice and plantations, and in the north and west where it is found in sessile oak wood and birch woods. Woodlarks like areas with short heathers and grasses and are one of the few birds well suited to clear felled plantation sites. Woodlarks nest and feed on the ground, mostly

on insects. During the winter months they spend more time in fields.

Sources: Fuller (1995); Peterson *et al.* (1983)

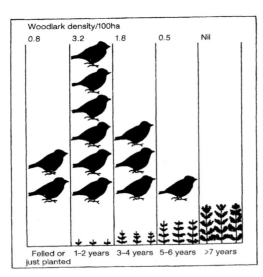

Figure 3.6: The density of woodlarks in young pine plantations of varying age in Thetford Forest. From Bowden and Hoblyn (1990).

Selection high forest

In this system, the trees are either felled as individuals or small groups (both selection felling) or larger groups of up to 0.1 hectare in size (group felling). The gaps are then filled either by natural regeneration or planting. The result is a mixture of different-aged trees with small gaps and small stands of similar-aged trees. The system is more difficult to manage than clear felling and care has to be taken when harvesting that the trees left are not damaged. It is also more difficult to plant up, or convert from existing forests. Selection by removing single trees is difficult to achieve in Britain because the tree species must be shade-tolerant, like beech, in order to develop under the canopies of the remaining trees, and mast years for beech are infrequent (Hart, 1991). Group selection is easier to manage and has the advantage of being perhaps the best of the high forest systems for nature conservation. This is because of the variety of tree age structure within a small area. Group selection may only create small clearings of less than 0.1 hectare so it does not necessarily have the same effect as coppicing a site in coupes. Although selection systems are commonly practised in continental Europe, they have not been very popular in Britain until recently.

Where selection felling removes single trees

or small groups it is sometimes called continuous cover forestry. Although it is standard practice in some European countries (e.g. Switzerland), it is not particularly popular in Britain. However, in recent years the Continuous Cover Forestry Group has been set up with the aim of encouraging the use of uneven-aged forestry systems and avoiding large clear fell areas (Helliwell, 1995).

A more formal version of this is a system devised by the sixth Earl of Bradford and his Forestry manager. The Earl divided his woodland into plots and planted Douglas Fir each year in prescribed plots. After growing for 18–20 years most of the original trees in each plot were removed leaving just one to mature. With a rotation time of 54 years, almost continuous tree cover was maintained and the result is a rather more organised group selection system (see Hart, 1991 for more details). In terms of nature conservation this system is very similar to the selection systems in the structure of the woodland but the very geometric design and single species management are disadvantages.

Shelterwood system

Here the trees are felled uniformly, in groups or strips across the stand. The aim is to encourage natural regeneration from the retained trees into the felled area. Once the regeneration is established another strip or block is felled. This system is better for conservation than clear felling but not as good as selection felling because there is a break in continuity when the last of the mature trees are felled (unless some are retained).

Regulations controlling tree felling

Trees can be subject to a range of regulations and in most situations the woodland owner, manager or tenant needs to obtain permission in order to fell them. A felling licence is needed from the Forestry Authority if it is intended to cut down more than 5 cubic metres of timber in a calendar quarter (e.g. 1st January to 31st March). No more than 2 cubic metres of this can be sold. There are various situations that are exempt, for example, pollarding, coppicing small diameter stems, felling dangerous trees and thinning small trees from forestry plantations, however, it is best to check first and the Forestry Authority publish a set of guidelines which can help. The felling licence must be received before any work is carried out and it often comes with a restocking condition, which means that new trees must be planted after felling. The Forestry Authority also gives grants in order to help with the cost of replanting.

Individual trees, and in some situations whole woods, may have Tree Preservation Orders (TPO) on them. Permission is needed from the local planning authority to carry out tree surgery work as well as felling.

Felling work in woodlands with a nature conservation status needs approval from English Nature, Countryside Council for Wales or Scottish Natural Heritage (depending on the country). In addition, if the work is likely to affect a Scheduled Ancient Monument then English Heritage, CADW (Welsh Historic Monuments) or Historic Scotland must grant permission. If the woodland does have some statutory designation, the best method of progressing is usually to have a management plan agreed by all interested parties. This does not negate the need for a Forest Authority Felling licence but it will make the progress of gaining permission easier.

Natural regeneration

Making use of naturally arising tree seedlings for the next generation of trees rather than planting has generally been less common in Britain than in other European countries. From the commercial point of view, planting is easier to manage, more reliable and it is easier to select specific strains and species. In terms of nature conservation, natural regeneration is usually preferable, especially if the woodland is ancient. It promotes local stock, genetic variation and a more natural stand diversity and species mix. However, some species regenerate better than others and sometimes undesirable species such as rhododendron and sycamore can become dominant; browsing by deer can also be a problem.

Harmer *et al.* (1997) studied natural regeneration in seventy-eight sites in southern England. Although they recorded tree seedling densities of over 10,000 per hectare, most were less than 0.2 m in height and about 30 per cent were browsed. The number of seedlings was lower than recommended by continental workers thus, while the density of seedlings may be acceptable for conservation and amenity interests, in most instances it is unlikely to be economic. However, John Workman's estate in the Cotwolds, which is predominantly beech, relies on natural regeneration (Hart, 1991). Both nature conservation practitioners (Peterken, 1993) and foresters (Hart, 1991) consider that more use should be made of regeneration in forestry situations. Certainly in terms of conservation it should be encouraged.

Coniferous vs deciduous trees

The extensive afforestation after the wars resulted in a backlash against coniferous trees, but is their reputation of dark and sinister woods with little wildlife justified? Scots pine is a native species (at least in northern areas) so, as expected, it does have a range of organisms associated with it. Some of these have a very restricted distribution and are found only within the natural range of pine. The bird fauna of native pine woods is generally richer than mature pine plantations in the same region (Fuller, 1995).

Exotic conifers and Scots pine in southern Britain do have specific faunas associated with them. In Chapter 2 the decomposer community of coniferous woodlands was briefly described but there are differences in other groups of organisms. Smart and Andrews (1985) point out that conifers benefit birds by providing evergreen winter shelter (in most cases) and they increase structural diversity and provide homes for birds specialising in conifers (e.g. goldcrest).

If possible it is usually better to have some native broadleaves in amongst the conifers. Bibby *et al.* (1985) found that 11 out of 22 species of breeding birds in conifer plantations preferred areas with broadleaved trees and that the overall density of these species increased with an increase in the broadleaves. Broadleaved trees allow the ground flora to be better developed and this probably increases the diversity of invertebrates (Peterken, 1996). The reverse is probably also true, some conifers in amongst broadleaves will probably also attract extra species.

Red squirrels (Species Box 3.10) prefer older stands of conifers and will sometimes thrive in plantations. Deer (especially red deer) are also able to live in commercial coniferous forests. Of the invertebrates, some are intolerant of coniferous plantations whereas others are more tolerant (Good *et al.* 1990), and some species emerge as pests (e.g. pine beauty moth and pine looper moth).

The coniferisation of broadleaved woods by planting up ancient and semi-natural areas is generally viewed as detrimental in terms of nature conservation. The growth of evergreen

Species Box 3.10: Red squirrel

Red squirrels are a uniform dark colour varying from deep brown to bright chestnut. They differ from the introduced grey squirrels in being smaller (180–240 mm body length, plus 140–195 mm tail) and in having ear tufts during the winter months. Red squirrels were once found in woodlands throughout the British Isles but started to decline during the eighteenth century. Despite some recoveries, in the early part of the twentieth century this squirrel became scarce while the introduced grey squirrel became more abundant. The reasons for the decline are still not clear but habitat destruction and disease are likely to have contributed. Direct impact by the grey squirrel is not now thought to have been a cause. In Britain today, the red squirrel is found in predominantly coniferous woodland, and in some places can co-exist with the grey. It is only found in deciduous woods where the grey squirrel is absent. The red squirrel is now found in the north of England, much of Scotland and most of Ireland. There are also a few outlying places where it is found such as the Isle of Wight and Thetford Forest in Norfolk. They feed mainly on seeds but take a wide range of other foods including fungi, fresh plant material and invertebrates. Pine cones are stripped of their seeds and scales leaving a recognisable core. Some food is cached during the autumn when seeds and fruits are plentiful. Usually mating takes place in the winter and three young are born sometime between February and April. If food is in short supply during the autumn months, breeding may be delayed so that the young are born later, between May and August. There may be two litters per year if there is plenty of food. Red squirrels build dreys in which to give birth. They are made of twigs and branches, usually at least 6 m up a tree, close to the trunk and often at a branch fork. Red squirrels do not hibernate during the winter.

Sources: Corbet and Harris (1991); Tittensor (1980)

conifers shades out the ground flora and thus has a detrimental effect on the animals feeding on such plants. Waring (1988) found a decline in the number of moths in coniferous plantations in relation to overgrown coppice but there were differences in the species recorded with some showing an increase in the conifer areas and others a decrease. The differences were largely attributable to the abundance of food plants. However, a few conifers in deciduous woodlands may increase the diversity of the site both in structure and by the addition of some specialist coniferous feeders. It is possible that some bats may find stands of conifers mixed with broadleaves beneficial (Holmes, 1997).

Brown (1997) points out that to many foresters the promotion of native species and discrimination against exotics appears extreme and reinforces the differences between forests for production and those primarily of interest for conservation. Commercial forestry may be the primary purpose of a particular area but it can also be sympathetic to wildlife. Likewise, a wood can have a very high conservation value but some timber can be removed for sale. If this helps the owner to be more positive about conservation it is likely to be beneficial in the long term.

Forestry and conservation

Malcolm (1997) notes that in the past the main objective of the forestry industry was to increase the total forest resource and to maximise the amount of timber produced. Now there is a shift towards high quality products rather than just quantity. Malcolm also notes that silvicultural systems need to be developed to ensure 'a transition from simple plantation forestry to properly structured and ecologically sustainable forests'. The Government published a report in 1994 called *Sustainable Forestry – the programme* which aims to set standards for the management of UK forests. There is an obligation put upon the Forestry Commission to achieve a reasonable balance between the interests of forestry and conservation (1985 amendment to the Wildlife and Countryside Act 1981). Since then the management for conservation has increased in importance in forests (Ferris-Kaan and Patterson, 1992) and one of the objectives outlined in the Forest Enterprise Corporate Plan (1997) is to enhance the nature conservation value of the national forest as a whole and to safeguard special habitats.

Guidelines for improving the nature conservation value of commercial forests

There are always some species that will thrive best in large clear felled areas and dense mono-specific blocks of conifers so it is impossible to give recommendations that suit every situation. The following are guidelines on how, in general, commercial forests can be made better for wildlife (from Fuller 1995 and Kirby 1984).

1 Native species are generally better than non-native.
2 A mixture of species is usually better than a monoculture.
3 Natural regeneration is better than planting.
4 If planting is necessary, use locally native species and local genetic types.
5 The edges are better sinuous and feathered, not straight and abrupt.
6 Make use of other existing features in the woodland (e.g. don't plant right up to the edges of streams).
7 Felling to produce a mixture of different-

aged trees is better than felling single-aged trees all at one time.

8 Avoid the use of herbicides and other chemicals if possible.

9 Retain some trees beyond the length of the rotation, especially damaged or misshapen ones with less timber value.

10 Design the woodland carefully (various guidelines are available, e.g. Forestry Commission 1991a).

11 Avoid planting conifers in ancient woodland.

12 Try to avoid extensive and repeated damage to rides during the course of the forestry work.

13 Where necessary, cleaning and pruning should be done carefully.

14 Felling small areas is better than felling large areas all at the same time.

15 Thin as early and heavily as possible.

16 The presence of a shrub layer is advantageous to many organisms.

17 Reduce the component of invasive non-native species (e.g. rhododendrons).

18 Wider spacing of the trees is better than very close (there is more scope for the ground flora to develop).

19 Minimal disturbance of the ground before planting is best. If some treatment is necessary, probably scarifying is best.

20 Dead or dying trees are best left to rot *in situ* if possible.

21 Cut branches from big trees are best left where they fall. Some brash is worth leaving.

22 If block felling is being carried out, it is probably best not to fell more than 10–20 per cent of the area in any 5–10 year period.

23 Avoid sensitive times of the year for harvesting and heavy work (e.g. not during March to July when birds are nesting).

24 If possible, have areas or networks of uncropped land where the trees are not managed.

25 Have a mixture of conifers and broad-leaved trees.

TO MANAGE OR NOT TO MANAGE?

From the previous sections in this chapter it might be thought that all British woods are actively managed as coppice, wood-pasture or commercial forests. This is by no means the case. Very many (and particularly those of a small size) are either not managed at all or receive occasional attention for a variety of reasons. There are estimated to be 175,000 hectares of small (under 10 hectares in size) neglected broadleaved woodland in Britain (Giles, 1996). Woods on estates or farms may be managed for game, small-scale fuel wood or, especially in upland areas, as an extension to the grazing land. Many owners such as local authorities, various trusts or private individuals do not have any resources to carry out substantial management even if they desired to. The result is that many British woods are managed in a rather haphazard way. The occasional tree is felled for timber, invasive species are cut back, a footpath is kept open, fly tipping removed, trees along roadsides trimmed, perhaps some bird boxes put up and the local children use it for cycling through. The diversity of management to which our woods are subjected is probably a good thing because it allows for regional variations and individuality. The major problem comes when a wood is to be felled to make room for houses, roads or farmland.

Minimum intervention

One final management option has to be considered further and that is taking the decision to do nothing. As non-managed woods in Britain are of a small size and usually subjected to outside influences (i.e. pollution and public pressure), true non-intervention akin to wildwood status is impossible to achieve. Minimum intervention is probably the closest we can achieve. Minimum intervention is not necessarily the result of no money or no will to manage, but can be a conscious decision. For example, it was decided in 1944 that Lady Park Wood in Gwent would not be managed in order that the natural processes taking place could be studied (see Peterken, 1996 and Peterken and Backmeroff, 1988 for details). Other woodlands probably do not need much active management in order to maintain them and limited intervention is often a good policy to employ in terms of nature conservation. It is also worth considering having minimum intervention areas within larger actively managed woods.

There are, however, a few potential hazards of not actively managing:

1 invasive and non-native species can spread unchecked and may be detrimental to the native flora and fauna, e.g. rhododendrons shade out the ground flora, they may be suitable for some birds to roost in but not much else;
2 grazing/browsing pressure from deer or rabbits can result in lack of tree regeneration for the future;
3 over time the tree composition may change and if the tree species is exotic the policy for that species may need to be considered;
4 non-intervention is not necessarily the best method of conserving the nature conservation interest of the site, this will depend on the species' present and past history (Read and Frater, 1993).

However, on the positive side, minimum intervention enables the study of ecological processes (Watkins, 1990) and can benefit the species of mature high forest. The best places for this form of management are large areas with little disturbance and where the structure is similar to high forest. Within actively managed woods, compartments can be left (e.g. coppices can be singled and then left) to provide more abundant dead wood and less disturbance of soils (Kirby, 1992 and Peterken, 1993). Evans and Barkham (1992) argue that non-intervention in coppiced woods will not result in reversion to their natural state because of the patchiness resulting from past management and human activities. They also point out that coppice may have been selected for specific ecological types among the plants, which may not survive if the area is left as minimum intervention.

RIDES, GLADES, FIREBREAKS AND EDGES

A variety of plants and animals thrive along woodland edges in glades and clearings within woods. These species probably existed in the wildwood, prior to extensive clearance, along edges formed by rivers and in tree fall glades, etc. They may also have been able to survive better under the woodland canopy because the temperature at that time is believed to have been warmer (Warren and Key, 1991). These species probably increased in abundance as woodland clearance proceeded and coppicing began but have declined more recently as traditional management has declined and the woods became fragmented (Peterken, 1996).

Rides and glades form bands and pockets of different habitat types within woods. They can be an important reserve for unimproved grassland (Peterken, 1996) but may also have characteristics of heathland, marshy areas and hedges. Open areas in woods can be sunnier and warmer and show a greater structural diversity, which suits some groups of organisms. Equally they can be drier, windier and allow light into the adjoining woodland (Hambler and Speight, 1995b), which can be detrimental to some groups.

Glades and rides in many woods do not appear to have any influence on bird distribution (Fuller, 1995), although the edges of woods are important for some species. Open spaces provide sheltered feeding areas for deer and good hunting grounds for bats. Some invertebrates such as butterflies, moths and flies favour clearings, for example, the speckled wood (Species Box 3.11). Even some of the true woodland invertebrates require nectar for the adult stages so gaps can be important to them.

One of the most studied groups with respect to glades and rides is the woodland butterflies

and some SSSIs have been notified for the butterfly populations of their rides (e.g. Bernwood, Buckinghamshire and Oxfordshire). Coniferisation itself has a detrimental effect on butterflies, but some large commercial plantations can be very good for butterflies because of their ride structure. Many butterfly species have been adversely affected by destruction of habitat and also the decline in traditional management such as coppicing and ride management (Hall, 1988). This is because most have food plants that are dependent on clearings both as larvae and the nectar-feeding adults. Steel (1988) studied the butterflies along rides and in glades at Sheephouse Wood, Buckinghamshire and recorded the following results.

1 Widened rides had more individuals and a bigger range of nectar plants for feeding than unwidened rides.

....

Species Box 3.11: Speckled wood

The speckled wood is a common woodland butterfly of central and southern Britain and Ireland. It is the British species most able to survive in dense woodland and exploits sunny patches. The adults can be found throughout the summer, from March to October and, like their habitat are dappled in colour. The males may behave in one of two ways. Either they perch in sunlit spots waiting for females to fly past, or they patrol up and down through their territory. There seems to be a slight difference in colour pattern on the wings depending on the behaviour the butterflies prefer. Perching males protect their territory from others; if another speckled wood flies through, the two butterflies spiral upwards towards the canopy and usually the original butterfly returns to the perch while the other one flies away. Females are also attracted to patches of sunlight but when approached by a male tend to drop to the ground where courtship

takes place. Mating follows in the tree canopy and eggs are laid on grasses, especially those in sheltered, isolated situations. The caterpillars are green in colour and feed on a range of grasses. The speckled wood is unusual in that the over-wintering phase can be either the caterpillar or the chrysalis.

Source: Thomas and Lewington (1991)

....

2 Widened rides with standard trees left were better for the true woodland species such as white admiral but generally had fewer individuals.

3 South-facing sheltered bays in the rides were probably more valuable than north-facing bays.

4 The ride with the highest number of species and individuals recorded was alongside a cleared area but 65 per cent of the butterflies were meadow browns and small skipper (both common grassland species).

These results illustrate some important points (see also Figure 3.7). Wide grassy rides may encourage large numbers of butterflies but if the species are very common grassland species and the opening up of the rides is detrimental to the true woodland species belonging to other groups, it may not be the best policy to persue.

Guidelines for the management of opens spaces in woods

Open spaces such as rides and glades can be managed to maximise their nature conservation value. The guidelines below are from P. Kirby (1992), Lane and Tait (1990), BTCV (1997) and Watkins (1990).

1 Know what species your management is aiming to encourage and be sure of their requirements. The following comments will apply if open spaces are considered appropriate.

2 Long-standing rides and clearings should be maintained as such if possible. Larger long-standing clearings should not be planted up. They may need to be grazed or mown/cut in order to be kept open.

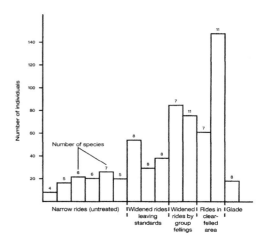

Figure 3.7: The number of butterfly individuals and species found in different types of rides in Sheephouse Wood. From Steel (1988).

3 Wider rides are better for light-demanding species than narrow ones. Ideally the width should be 1.5 times the height of the trees along the edge to ensure that sunlight reaches the ground at all times of the day. At least 9 m wide is best for fire breaks.

4 The edges are better if they are sinuous, forming scallops rather than being straight. Scallops of 30×20 m are good for butterflies but smaller ones are better than nothing. Having branches hanging over the rides in places is fine.

5 Curving rides are better than straight ones as they are less likely to become wind tunnels.

6 Constricting or closing off the ends of the rides at the edges of woods can help to stop spray drift, etc. and the wind.

7 Vary the orientation of the rides if possible. East–west is generally best especially in the summer months as this orientation gets the most sun. North–south rides are useful for some species as they always get some light in the winter months. They do not all have to be wide and open.

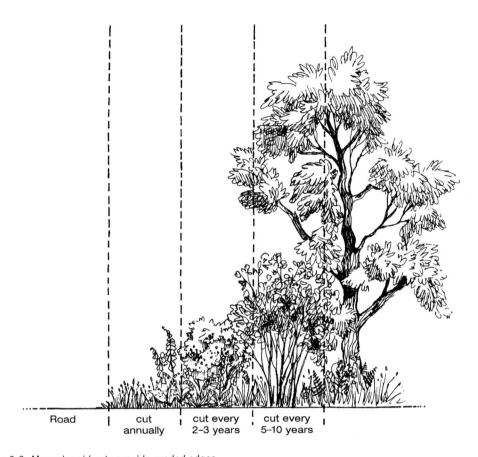

Road | cut
annually | cut every
2–3 years | cut every
5–10 years

Figure 3.8: Managing rides to provide graded edges.

8 Be careful not to destroy old hedges, ditches or wood banks and follow ancient patterns of rides and tracks if possible.

9 Orientate the rides so they are not ideal for the prevailing winds to howl down them and make them cold.

10 Aim for a graded edge structure from short grass in the middle of the ride to longer nearer the trees (see Figure 3.8). Maintain this structure by cutting if necessary.

11 If cutting the edges, use a rotation. Cut adjacent and opposite patches in different years so a mosaic of different ages is formed. Aim to cut just a small patch each time.

12 If mowing rides, aim to remove all the cut-

tings or leave a few large piles of cuttings. Ideally mow after seed has set (September–March).

13 To maintain the edges of the rides consider cutting coppice. This is better done as a wide strip rather than a very narrow one and is best done in winter.

14 Bare ground in the centre of rides is good for invertebrates such as bees and wasps. Wet ruts are also good but it is best to avoid continual disturbance.

15 If creating rides, try to restore old ones if possible and do not destroy higher nature conservation interest (e.g. shade-tolerant species) in creating them.

16 Glades can be created at intersections of

rides and this helps mobile species to use an inter-linked system.

17 Glades are ideally at least 0.1 ha in size with a well-graded edge or 0.2 ha if surrounded by tall trees.

18 Longer-lived glades are usually more valuable than temporary ones.

19 Some glades may remain open without management if grazed or bracken-dominated but they may need management similar to rides. Coarse vegetation (e.g. brambles) may need to be kept in check.

The management of rides and associated habitats is especially important in commercial plantations as they may be the most ecologically valuable part of the site.

Many species use the edges of woodlands, which can be managed like rides; a gradual transition and a scalloped edge being better than an abrupt straight edge.

GRAZING IN WOODLANDS

True natural woodlands with no human intervention would have contained large herbivores. These would have grazed the ground vegetation in clearings and browsed the lower branches of some of the trees and saplings. Today the large herbivores grazing in British woodlands differ in species and abundance from those wildwood times and take two forms. First, domestic stock (Plate 3.6), usually cattle, sheep or horses and second, deer. The woodlands they graze range from wood pasture and parklands to upland sessile oak woods and coniferous plantations. It is likely that low-intensity grazing is beneficial in most woods, it keeps some plants in check, enabling a variety of ground flora to survive but does not hinder natural regeneration of the trees. Grazing at low levels may even stimulate the natural regeneration of the trees. However, as grazing pressure increases, the field and shrub layers gradually reduce in height and tree regeneration declines (Table 3.7), grasses increase and broadleaved herbs decrease. Because of the effects of grazing on the plants, other groups of organisms are affected. Birds favouring structural diversity (e.g. warblers) decline but a few species increase in

Plate 3.6: Woodland grazing in the New Forest, Hampshire. Note the closely cropped holly at the base of the tree.

Table 3.7: The effects of increasing grazing pressure on woodland organisms

	No grazing					→ High intensity grazing
Trees and shrubs	No regeneration due to high competition	Creation of regeneration niches	Loss of seedlings. Damage to seedlings	Loss of saplings. Severe tree browsing	Barking of mature trees. Loss of shrub layer	Creation of parkland, moorland or heathland
Higher plants	Reduced diversity, dominated by a few vigorous species	Reduction in vigorous species. Increase in diversity	Reduction in vegetation structure. Increase in grazing tolerant species	Loss of plant diversity, particularly grazing sensitive species	Loss of cover and damage due to trampling. Bare ground	Impoverishment due to net loss of nutrients from the system
Lower plants	Reduced cover and diversity due to competition from higher plants	Increase in cover of ground dwelling species as competition from higher plants reduced		Damage to ground dwelling species due to trampling	Reduction of drought sensitive bryophytes	Increase in epiphytic lichens associated with parkland
Small mammals	High small mammal populations, a few species predominate	Increase in diversity as structural diversity increases	Reduction in small mammal populations as ground vegetation structure simplified		Reductions of populations through competition for food	Loss of diversity and abundance. Species of open ground predominate
Birds	Favouring birds of dense shrub layers. Low numbers of ground nesting species	Increase in diversity as structural diversity increases	Increase in species favouring low shrub cover	Loss of ground nesting birds due to poor concealment	Loss of species dependent on berry bearing shrubs	Reduction in raptors dependent on small mammal populations
Invertebrates	High populations of plant living species	Increase in diversity as sward structure diversified	Increase in dung using species	Decline in woodland species		Increase in parkland, moorland or heathland species

Source: Mitchell and Kirby (1990) with additions from Fuller (1995)

Note: Shaded columns indicate the optimum grazing intensity for nature conservation purposes.

compensation (Fuller, 1995). Small mammal numbers decline which leads to a corresponding decline in mammalian predators (e.g. stoat and weasel) and avian ones (tawny owls and hawks) (Tubbs, 1997).

The open aspect of heavily grazed woods may favour lichens, bryophytes and, if the trees are old or pollarded, saproxylic invertebrates. However, other invertebrates can decline in numbers. The nectar sources in the shrub and field layers decrease so nectar-feeding invertebrates such as butterflies follow suit. Perhaps the biggest concern of heavy grazing is the lack of tree regeneration. Figure 3.9 illustrates the animal numbers grazed in the New Forest and the major periods of tree regeneration and it can be seen that they are very closely linked. In an area as large as the New Forest (19,771 hectares, Tubbs (1997)), successful regeneration does take place even during periods of high grazing pressure, where fallen branches prevent grazing and in holly patches (Peterken, 1993), although this may not be true of smaller woods.

A total lack of grazing may not be ideal either (Kirby et al., 1994). This can lead to

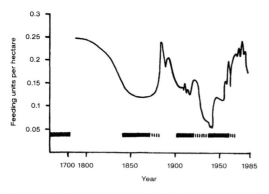

Figure 3.9: Tree regeneration in the New Forest in relation to the number of grazing animals. The black bars indicate the main periods of tree regeneration. One feeding unit equals one pony, 0.5 cattle or 0.34 fallow deer. From Tubbs (1986).

dense tree regeneration, which can cause loss of lichens and bryophytes, but keeping grazing to ideal levels is very difficult. Some wood-pasture is grazed with stock for economic or aesthetic reasons and upland woodlands provide valuable shelter for sheep especially in the winter months. Deer numbers are high in many parts of Britain now and are likely to continue to increase (Watkins, 1990), it is expensive to exclude them. For woods that are heavily grazed, the best solution seems to be either to try to exclude herbivores by fencing them from the entire wood or by reducing the numbers from sections of it in turn. Alternatively, in the case of deer, by controlling numbers in the local area which is usually only successful if carried out by neighbouring land owners (but see case study on Caledonian pine, p. 126).

To encourage regeneration, total exclusion may not be necessary but if low densities of grazers are not achievable, then periods of exclusion from parts of the wood for long enough for the young trees to grow out of danger should be aimed for. The only exceptions to the 'light grazing' rule might be woodland that has been grazed for long periods (e.g. the Royal Forests and medieval parks). Their ecology has developed over hundreds of years and the landscapes are historically important. But even here care needs to be taken to ensure the next generation of trees can establish. Kirby et al. (1994) stress the importance of having long-term objectives and planning for the future with regard to the grazing of woods.

NEW WOODLANDS

Afforestation is continuing in upland areas, especially in Scotland, primarily for the purpose of commercial forestry and for a long time planting was largely restricted to the

Plate 5: *Pyrochroa serraticornis*, a red-headed cardinal beetle, the larvae of which live under wet loose bark of a variety of tree species

Plate 6: Horse extracting timber, Cumbria

Plate 7: Old growth Scots pine forest, Rothiemurchus

Plate 8: An Atlantic oak wood, near Ffestiniog

uplands. However, in recent years, more interest has been taken in planting in the lowlands and in the use of broadleaves and mixtures rather than just conifers.

In 1995 the government drew up a White Paper called *Rural England*. It stated that the government 'would like to see a doubling of woodland in England over the next half century'. However, now the government has changed, this is no longer an official target. To double the amount of woodland would mean planting 1 million hectares to give 15 per cent woodland cover, the same as at the time of the Domesday Book. The discussion document subsequently produced (Forestry Commission/ Countryside Commission, 1996) lists ten reasons for creating and managing new woods.

1 To provide timber and other wood products.
2 To provide opportunities for sporting and recreational activities, and space for those seeking a place to relax.
3 To enhance the beauty and character of the countryside and contribute to maintaining the diversity of rural landscapes.
4 To help revitalise derelict and degraded land.
5 To enhance wildlife habitats.
6 To create jobs and provide opportunities for economic diversification in rural areas.
7 To provide a link with the past and help create a distinctive local identity.
8 To be a valuable educational resource, particularly for ecological study and physical education.
9 To improve the quality of life in and around towns and cities.
10 To contribute to reducing atmospheric concentrations of carbon dioxide, the main gas responsible for global warming, by absorbing and storing carbon.

The main functions of new woods are likely to be threefold: for commercial and economic reasons; for social and recreational reasons; and for environmental reasons. Already the National Forest in the Midlands and twelve Community Forests have been started. The Community Forests are being promoted by the Countryside Commission and Forestry Commission and are close to urban fringes. They will consist not only of woodland but will include open space and farmland. These two initiatives alone will result in 80,000 hectares of new woods (Forestry Commission/ Countryside Commission, 1996). These forests are aimed at multipurpose use, i.e. the three functions given above.

Where to locate new woods?

There are many factors that are likely to influence the site of a new wood, but the golden rule is not to plant on an area which currently has a higher conservation value that the woodland will create. Examples include heathland, unimproved grassland and land used by nesting waders (curlew, snipe, etc.). Good places to plant new woods include those where they will expand or link existing woods, where they will provide green corridors and link existing appropriate conservation sites and, if recreation is an aim, areas close to centres of population.

What to plant?

Two main types of woodland can be considered. First, woodland primarily for commercial forestry, where a good commercial crop is the main aim. Exotic coniferous species may be the most appropriate, although some introduced broadleaves, such as sweet chestnut, can also be used. Where possible, native species should be encouraged in the fringe areas (e.g. edges and along rides). The second type of woodland is where only native species are planted which is a better option if nature conservation is the main aim. However, these woods are likely to produce a lower economic return than exotic species. If the latter option is chosen, the most appropriate species can be selected, using those abundant in the local area and best suited to the soil and climate. Rodwell and Patterson (1994) give details of how to use the National Vegetation Classification to identify appropriate species and also give comprehensive guidelines of how to plant a new wood.

How to plant?

The design of the new wood will depend on its main function. Ease of access for management is paramount in commercial forests whereas aesthetic considerations may be more important if the wood is to have a major function as a recreational area. The general appearance in the landscape is an important factor, whatever the future function of the wood. A Forestry Commission (1991a) booklet covers these aspects of designing new woods as well as the internal structure.

One important point is that trees can be relatively easily planted but the field layer plants take many years to colonise and are expensive to plant.

Genetic provenance

Awareness is growing of the importance of using locally native plant stock for planting, not just native species. Amongst the commercial forestry industry, selection of tree stock has, in the past, been primarily on the basis of characteristics suitable for good timber production. There is increasing realisation of the importance of keeping a wide genetic base for research and development of new strains. In addition, there are several instances where imported strains have proved unsuitable for British conditions (Malcolm, 1997).

Maintaining local strains of species within different parts of Britain keeps the genetic variety and is especially important in woodlands that are nature reserves. Wherever possible in ancient woodland any planting should be of local stock. Considerable success has been obtained with such projects, for example, at Fellbrigg Hall in Norfolk (Battell, 1996). The organisation Plantlife, in conjunction with other bodies, has recently launched a project called Flora Locale to help promote the use of indigenous plants from the local area.

RECREATION IN WOODS

Increasing use is being made of woodlands for recreation, which does raise some important issues with regard to management. Figure 3.10 shows the number of people per hectare of woodland in England. Perhaps not surprisingly the highest numbers are close to urban areas but other regions may be popular holiday resorts and have large numbers of seasonal visitors.

The largest land manager in the UK is Forest Enterprise, with 37 per cent of the country's woods and forests. One of their objectives is 'to encourage access on foot and to develop the recreational potential of the estate' (Forestry Enterprise, Corporate Plan, 1997–2000). In order to do this they provide and maintain visitor centres, holiday accommodation and over 500 picnic sites (Forestry Commission, 1997) for their estimated 50 million day visits a year. Forest Enterprise is the third largest

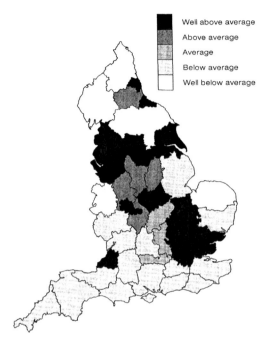

Well above average
Above average
Average
Below average
Well below average

Figure 3.10: The number of people in England per hectare of woodland. From Forestry Commission/ Countryside Commission (1996).

Plate 3.7: A forest campsite, Glen Moore

manager of camping and caravan sites (Plate 3.7) and the largest manager of countryside recreation facilities in the UK and has a policy of freedom of access on foot where possible. Other woodland owners have similar policies, for example, the Woodland Trust which owns 5,260 hectares of woodland, has open access on 95 per cent of land (Countryside Commission, 1993b). Access to other woods is variable, often depending on the owner of the land, but it is likely that a considerable amount of the new woodland being planted will have public access through the community woodland policies. The Countryside Commission is continuing to encourage woodland owners (especially local authorities) to open woods to the public (Countryside Commission, 1993b).

Woodlands can absorb large numbers of people without appearing crowded. For woods over 75 hectares it can be possible for over 100,000 visitors a year to come and still obtain a sense of seclusion (Forestry Commission, 1991a). While recreation and commercial forestry can, with care, be accommodated side by side relatively easily, the integration of large numbers of visitors and a high ecological value can be more difficult. In certain situations it may be possible to operate a zoning system so that more intensive recreational activities are permitted in some places whereas more sensitive areas have limited access to visitors.

Recreation includes a wide range of activities from walking and horse riding (Plate 3.8), to war games and off-road driving. The impact on wildlife can be as direct effects, like dis-

Plate 3.8: A forest being used for recreation, the New Forest, Hampshire

turbance of animals by the presence of people and trampling of vegetation, or indirect effects such as lack of suitable habitat or food for animals because of the lack of vegetation. People also bring other problems in the form of dogs which defecate (and alter the nutrient levels of the soil) and chase woodland mammals and ground nesting birds. Visitors may also leave behind litter, which can kill mammals or result in fires. It is extremely difficult to measure the impact of recreational pressures on woods and very little research has been done. There are some pieces of evidence indicating that disturbance may adversely effect birds (Fuller, 1995) and in instances where the ground layer under trees is totally denuded by visitors (especially in small woods) there is clearly some effect.

Visitors may be invited into the woods in order to obtain revenue (either directly or by making use of other facilities); in order to increase public understanding; for educational reasons; to involve the local community; for good public relations; or to control where they go more easily (Forestry Authority, 1992). When visitors are encouraged, the woodland may be managed in a slightly different way.

Most visitors feel safer with well-defined routes and paths but these must be maintained once provided. On the positive side, they can be used to divert people from sensitive areas.

One important consequence of inviting visitors into a woodland is the health and safety aspect. In woodland settings the potential hazard of a person or vehicle being hit by a falling branch is obviously greater than in the open. The landowner must take 'every reasonable precaution' to avoid situations which might endanger visitors. This means that trees close to car parks, picnic sites, etc. should be inspected regularly and any necessary work done to make them safe.

The educational value of woodlands is immense for both children and adults, as trees are a very emotive subject. It is gratifying that people are concerned about the countryside and are increasingly being invited to play an active role in planning and managing woods. However, in comparison to woods, people are transient; the Forestry Commission (1997) notes that the average household now moves every seven years, whereas the majority of trees take at least two generations to mature.

MANAGEMENT FOR GAME

It has already been mentioned that coppicing encourages deer and this can be used to advantage if shooting is important for the income of a woodland. Unfortunately Ratcliffe (1992) shows that the numbers of deer ideal for sporting activities is very close to that where a heavy impact can be seen on broadleaved woodlands.

Pheasant shooting is a more widespread sport than deer stalking. A questionnaire of Country Landowner Association members showed that 67 per cent had retained existing woodlands because of the game interest. Pheasants spend the spring and summer feeding on growing crops but in the winter months spend much of their time in woodlands. The territories of the males include woodland edges and the birds rarely penetrate more that 50 m into the wood (Bealey and Robertson, 1992). For this reason coppiced woods are often ideal. The pheasant holding capacity of a wood depends upon the amount of edge and quantity of shrubby cover (Hampshire County Council, 1991).

Lane and Tait (1990) list the woodlands most suitable for pheasants (in descending order of preference) as follows:

1 managed traditional coppice;
2 young mixed conifer/broadleaved woodland;
3 broadleaved woodland with an open canopy and shrub layer;
4 mature broadleaved woodland with closed canopy and bare floor;
5 mature coniferous woodland with closed canopy and bare floor.

The methods used for pheasant rearing and shooting are not always ideal in terms of the nature conservation value of the woodland but usually this can be overcome. Carroll and Robertson (1997) provide a useful guide to the integration of pheasant management and conservation.

THE PRINCIPLES OF WOODLAND MANAGEMENT

Peterken (1993) provides fifteen principles, primarily from a conservation point of view, which give a useful summary of this chapter.

1 Distinguish between:
  (a) individual woods of high conservation value;
  (b) woodland areas of high conservation value; and
  (c) other woodlands.
2 Afford special treatment to special sites and special areas.
3 Minimise clearance. Necessary clearance should avoid sites of high conservation value.
4 Accept afforestation, except on sites of high conservation value, but not so much, that non-woodland habitats are reduced to small islands.
5 Develop (or retain) large blocks of connected woodland, while maintaining a scatter of small woods between large blocks.
6 Minimise rates of change within woods.
7 Encourage maturity by maintaining long rotations. If this is not possible, retain a scatter of old trees after restocking.
8 Encourage native tree species and use non-native tree species only where necessary.
9 Encourage diversity of
  (a) structure;
  (b) tree and shrub species; and
  (c) habitat in so far as this is compatible with other principles.

10 Encourage restocking by natural regeneration or coppice growth.
11 Take special measures where they are necessary to maintain populations of rare and local species.
12 Retain records of management.
13 Manage a proportion of woods on non-intervention lines in order to restore natural woodland in so far as this is possible.
14 Maintain or restore traditional management where this is possible and appropriate.
15 Where traditional management is not possible or appropriate, introduce alternative systems of management, which retain or enhance the conservation value of special sites and areas.

English Nature (1998) has produced a set of guidelines presenting the various options for managing ancient woodland.

THE FUTURE

While the traditional management systems of medieval times can still be seen in the UK, over the years they are gradually becoming adapted. No longer is the production of timber or underwood the sole purpose and today conservation and recreation are as important. As the functions of many woodlands become multipurpose, this will inevitably have an impact on the landscape. The angular outlines of commercial plantations will gradually soften, new woodlands will be planted and there will be more pressure on woods of all sizes and types to be economic and 'sustainable'.

While there is no denying that public access and the production of timber are important, we must not lose sight of the fact that British woodlands hold a high proportion of the country's biodiversity and we have an obligation to look after this resource. It is perhaps more difficult to plan for the future management of woodlands than any other habitat because trees live such a long time, human memory is short and results are sought quickly.

4

CASE STUDIES

•

The following case studies expand the wood-land management section by presenting specific examples of woods and outlining details and problems from woodlands throughout the UK (Figure 4.1). While they provide examples of work being carried out, each wood is unique and management decisions have to be taken on an individual site basis.

BURNHAM BEECHES, BUCKINGHAMSHIRE

Burnham Beeches is an example of an historic wooded common. For much of its history most of its 220 hectares were grazed. The southern part of the site was heathland with wet boggy areas in lower lying places; to the north was 80 hectares of wood-pasture (Figure 4.2). The most important part of the wood-pasture was the pollarded trees, most of which were beech with some oak. The grazing land under the trees was probably a mixture of acid grassland and heathland. The decline in traditional management practices on the site has had two major consequences. First, the decline in grazing, leading to a total lack of animals since the Second World War, has resulted in the infilling of the open areas between the ancient pollards by secondary birch-dominated woodland. Second, the pollarded trees have not been cut for over 150 years. The pollards themselves are estimated to be between 400 and 450 years old and most are either hollow or rotten. The weight of the

Figure 4.1: Map showing the location of the case studies mentioned in the text.

large branches resulting from the lapse in pollarding has caused many of the trees to fall over or split apart (Plate 4.1). From an estimated maximum of 3,000 pollards there

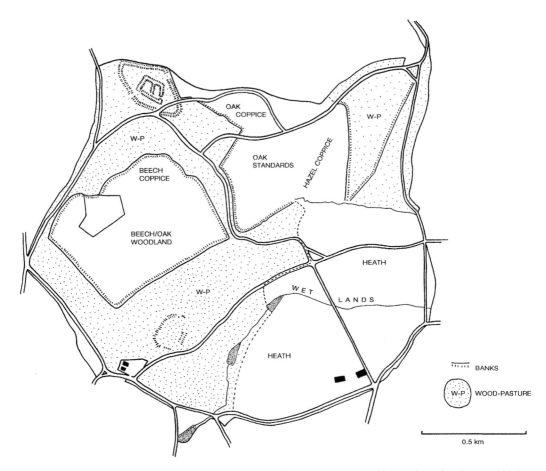

Figure 4.2: Map of Burnham Beeches showing areas of different management history. From Read *et al.* (1996).

are now less than 540 of the old generation left.

Burnham Beeches is a National Nature Reserve primarily because of the old trees and their associated saproxylic fauna, thus abundant dead wood both on the ground and in the trees themselves is of primary importance. Current management aims to secure the continuity of this rare resource.

Up until a few years ago there were very few places in Britain where the practice of pollarding trees other than willow or street trees was carried out on a regular basis. More difficult is the consideration of cutting trees that have not

been cut for over 150 years (when the normal cycle was probably 10–20 years). Beech is also an especially notorious species as it was thought to rarely respond to cutting. Rather than attempting to re-pollard the veteran trees first, experiments were carried out on young beech maidens to ascertain the best cutting techniques. The main lessons learnt were to ensure the trees had adequate light (although opening them up quickly from dense shade can result in sun scorch and drying out of the trunk). More importantly, it is necessary to achieve a balance between taking off enough branches to encourage the tree to put on new

Plate 4.1: A wind-blown beech pollard, Burnham Beeches, Buckinghamshire

Plate 4.2: Cutting a beech pollard again. This tree has not been cut for over 150 years, Burnham Beeches

growth without removing so many that it dies. Work started on the old generation of pollards in 1989 and consists of removing competing tree growth (mostly birch and holly) and reducing the crown of the trees so that they are more stable, while still leaving plenty of canopy (Plate 4.2). Tree shape is immensely variable so that some are easier to cut than others. New pollards are being created alongside the older ones to provide a new generation. It has been found that even quite old beech trees (over 100 years) will pollard successfully as long as it is possible to leave enough, preferably a complete ring of existing branches, having completed the pollarding operation.

An experimental plot of 6 hectares of the former wood pasture site was fenced and subsequently grazed following the clearance of secondary woodland in 1989. The grazing is poor in quality and much of it is closer to heathland than grassland. In recent years Exmoor ponies, British White cows and Jacob sheep have grazed the area. Berkshire pigs were turned out during the autumn months in the first few years after clearance. They turn over the humus layer and top-soil through rooting, this assists regeneration and also helps to control bracken, which spreads in the open areas, by the pigs eating the underground rhizomes. Traditionally pigs were turned out at this time of the year, known as the pannage season, to fatten up on beech mast, acorns and other autumnal fruits. The small area of wood pasture grazed at the moment probably cannot sustain pigs every year and allow a sward to develop. The ponies and cattle also graze 32 hectares of heathland on the site and the pigs have been turned out here in recent years. The sheep have proved slightly more effective than the other animals in keeping down the growth of young birch trees but also browse some of the plants that are being encouraged, e.g. heather.

The work plan is to continue the pollarding across the whole of the wood-pasture area. The old beech trees are being lost at an alarming rate and the cutting technique does seem to prolong their life by relieving the weight of the branches on the hollow stems. At present no other areas are grazed except the experimental plot of new young pollards.

Monitoring of the ground vegetation and ground-running invertebrates is carried out in some of the actively managed areas and all work carried out on the trees is recorded. A considerable amount of survey work has also been done in recent years so that the rare and important species are relatively well documented.

An added dimension to the site is that it was purchased, and is still managed and maintained, by the Corporation of London under their 1878 Act of Parliament. The original purpose of the purchase was to provide a rural open space for the recreation of the people of London. Visitors today come from the local community as well as from across the world and number between 0.5 and 0.75 million per year. An increasing problem with large numbers of visitors on sites such as this is the risk of litigation should an accident occur due to a branch falling on a vehicle or person (see Chapter 3). At Burnham Beeches a system is in operation whereby high priority areas are assessed regularly and any necessary work

done to make trees in these areas safe. Wherever possible, remedial tree surgery work is done rather than felling at ground level. This regular monitoring of potential hazards is essential in fulfilling the obligation of 'reasonable care' placed on owners in public liability situations.

BRADFIELD WOODS, SUFFOLK

Bradfield Woods are an outstanding example of a working 'coppice with standards' woodland (Plate 4.3). The history of the woods is well documented and, apart from a lapse of just thirty years, the coppice-style management has continued since at least 1252 (Rackham, 1990). The woods belonged to the Abbey of Bury St Edmunds and originally were in two parts. Felshamhall Woods were predominantly coppice, while the adjoining Monk's Park was a fenced deer park. In the 1960s part of Monk's Park was destroyed and the remaining area is now mostly coppiced. Since the 1970s 70 hectares of the woods have been owned by the Wildlife Trusts and are now managed by the Suffolk Wildlife Trust.

When they were acquired the decision was made to continue coppicing. This was for two main reasons. First, the completeness of the estate and the corresponding records mean that for purely historic and cultural reasons this site is a valuable working woodland (see Rackham, 1990 for more details). Second, the woodlands are valuable for a range of animals and plants which are traditional coppice specialists, for example the ground flora includes oxlips, wood anemones and wood spurge. Common dormice are also found although the numbers are probably quite low. The structure of the coppice also suits many summer migrant birds and the woods are thus an extremely important local site.

Plate 4.3:
Coppice at
Bradfield Woods,
Suffolk

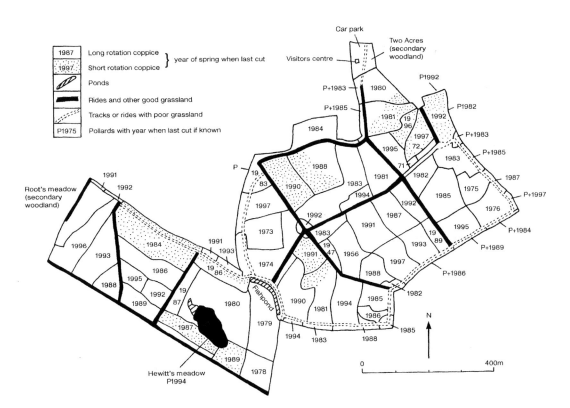

Figure 4.3: Map of Bradfield woods showing when blocks of coppice were last cut. From Archer *et al.* (1995).

Currently there are two different coppice cycles cut (see Figure 4.3). One is a short rotation of 8–12 years and takes place in the hazel-rich areas. The coppice is cut by a local hurdle maker who takes the hazel for his own use and leaves the other underwood species. The longer cycle of 20–25 years takes in areas where a much greater range of species are cut including ash, wych elm, oak, birch, alder, field maple and lime. The products sold from the wood include firewood, poles of various sizes, fencing stakes, hazel for thatching spars, and wood ash for pottery glazes. Most of the firewood is sold within 6.5 km of the reserve. Until recently much of the produce from the coppicing went to the local rake factory. This has now unfortunately closed down. The aim is to maximise income but in all aspects nature conservation takes first priority.

The standards in the coppice plots are mainly oak with a few other species interspersed. Many standards were felled in 1929 so most of those now left are rather smaller trees than might be expected. The aim is for the density of standards to produce a cover of 10–15 per cent. Natural regeneration is very good (except for the lime which only suckers) so there is always a range of new, young trees.

One increasing problem is the number of deer (roe, fallow and muntjac) and these cause problems by eating the coppice regrowth. Currently, the freshly cut coupes are fenced using the small brash from the coppicing which is of limited economic use. The brash is first heaped up, then stakes put through it and the tops woven. These 'dead hedges' are very sophisticated (Plate 4.4) and work well, keeping out most of the deer.

Bradfield Woods are among the best woodlands in Britain in terms of the number of plant species (they probably also have the highest number of ancient woodland indicator species). This is partly because of the very variable soil types. While the area is generally basic and waterlogged, there are patches of well-drained sand and loess which are glacial deposits from the Ice Age. These more acidic patches tend to be bracken glades. Scattered in the woods are also wet hollows where alders predominate.

Plate 4.4: Dead hedging, Bradfield Woods, Suffolk

Some of the associated habitats on the site are as important as the coppice. In particular the woodland rides are especially good for plants and provide sunny and sheltered places for invertebrates. A number of plant species are found predominantly in the rides rather than in the coppice plots. Some of the rides get disturbed in the winter months by vehicles. To help protect them, short lengths of hazel have been laid into the tracks. This works well and is in keeping with the visual appearance of the woods. Hewitt's meadow is all that remains of the wood-pasture of Monk's Park. This glade-like meadow acts as an extension to the rides and suits a range of grassland species. There are also two ponds, one ancient and one recent, which, along with the surrounding ditches, are significant water features providing good habitats for dragonflies and other aquatic species.

As the site has been a working coppice for a long time, there is little in the way of dead wood and very few saproxylic species. The exceptions to this are the pollarded oak and ash trees round the boundary of the woods. These are now also being cut again and some new pollards started. Recently the management has included retaining dead standing trees and also dead poles on coppice stools to increase the dead wood component. In addition, holes have been bored into some of the poor quality oaks to encourage heart rot. Some small parts of the woods have been designated non- or minimum intervention areas and are no longer actively managed.

The woods have free access and are used greatly by schools and the general public. Bradfield has been promoted as one of the Suffolk Wildlife Trusts' 'showpiece' reserves so the educational value of the woods is likely to increase in coming years.

KIELDER, NORTHUMBERLAND

Britain's largest commercial forest is owned and managed by Forest Enterprise. At 62,000 hectares it produces 5 per cent of all the timber grown in Britain but it is much more than just a commercial forest. The first trees were planted at Kielder in 1926 at Smales, on 800 hectares. Subsequently 19,000 hectares were purchased by the Forestry Commission in 1932 and the estate continued to expand until 1969. By the early 1970s over 50,000 hectares of land had been planted up (Plate 4.5).

The climate and soils in the area limit what can be grown at Kielder and initial plantings were all coniferous species which were able to survive and meet the objectives at the time of planting, i.e. producing a high volume of timber in a short time. The original aim of Kielder, solely that of timber production, has changed over the years. The management objectives are now:

1 to provide a steady and sustainable supply of timber;
2 to encourage the public to use the forest as a recreational resource by providing appropriate services;
3 to sustain and enhance wildlife conservation;
4 to ensure that forest management operations are carried out in such a way that the forest develops into a positive feature of the landscape;
5 to generate income to pay for all these activities and provide a financial return on investment.

Timber production

The main timber species is Sitka spruce (Species Box 4.1). Norway spruce grows

Plate 4.5:
Commercial forestry,
Kielder,
Northumberland

reasonably well but other conifers are not so successful. The forest was planted over a relatively short period of time (over half was planted between 1945 and 1960). Therefore much of it will also mature and be ready for harvesting at the same time. The original planting was blanket-like, with large blocks in geometric shapes. This is not aesthetically pleasing and is not so good for nature conservation. Consequently, there has been much time and effort recently put into altering the structure and aiming for a patchwork of different tree ages and more sympathetic outlines. Because of the longevity of the trees this inevitably takes time and so far about half of the forest has been redesigned, working from the north east to the south west. This redesigning does incur financial losses.

Very little thinning is carried out in Kielder. This is because most of the forest is very exposed, has poor soil structure and the trees are very prone to being blown over. Thus most of the trees are clear felled on a shortened rotation before they reach a height where they are susceptible to strong winds. However, 10,000 cubic metres of thinnings per year are produced, from certain areas, and this is done in one of two ways.

1 As design felling. The aim of this is to lengthen the cycle between felling coupes and to break up the age structure. This work is done with machinery.
2 With horses where the work is needed to be very sympathetic (e.g. round Kielder Water, and the A68 at Redesdale). Approximately 60 trees per day are felled using chain saws and extracted in this way. Some mature timber is retained, especially Norway Spruce and Scots pine, as it is good for red squirrels.

The trees are mostly harvested at between 35 and 55 years old with some being felled earlier or left longer as part of the forest redesign programme. Each year 1,100 hectares are felled which amounts to about 1,500,000 trees, and this produces 400,000 cubic metres of timber. Thus 1,000 tonnes of timber, approximately 50 lorry loads, are

Species Box 4.1: Sitka spruce

Sitka spruce was introduced to the UK from the west coast of North America. Probably most trees grown here now originate from the Queen Charlotte Islands. Sitka spruce has proved to be a very valuable timber species as it is reasonably frost hardy and can tolerate very exposed situations. It does less well on very dry sites, waterlogged ground or very poor soil unless nutrients are added. Sitka spruce is now the most commonly planted species in commercial forests as it is fast growing (over 1 m per year on good sites) and has good stem form. It can also withstand higher levels of deer browsing than other species but does suffer from green spruce aphid although it usually recovers. It is best grown in areas where the annual rainfall exceeds 1,000 mm and where there is not too much frost. The needles are strong and pointed with a ridge on the underside. They are green above and paler whitish beneath. The new foliage appears quite bluish in colour. The twigs have numerous side shoots and the remains of old leaves that remain as small pegs. The flowers are hard to find as they occur high in the canopy. The female flowers start as tiny pink cones, these develop into pale, papery, narrow cones about 70 mm long. The seeds are very

small with long wings. The wood is light in colour and weight. It is easily worked and can be used for a variety of uses but most goes for the pulp and paper industry.

Sources: Hart (1991); White (1995)

harvested every working day. Each tree is approximately 0.2 m³. The quantity of timber produced has now reached a plateau and may drop slightly in future years. Of the trees felled, 50 per cent is sold standing and 50 per cent is dealt with in-house. The future aim is for 60 per cent to be sold standing and 40 per cent to be harvested by Forest Enterprise staff.

The trees sold standing are either paid for on a tariff system or over a weigh-bridge. A large auction is usually held in February with three to four additional smaller sales throughout the year. A small amount of timber is sold through the electronic auction system. The auctioneers hold the details of the timber and

bidders offer prices for it by telephone. The felling is then carried out by the companies who win the tenders, all of which is done mechanically. A single man with a mechanical harvester can fell, remove the branches and cut to length 4–500 tonnes per week.

Current (1998) timber prices per tonne (at roadside) are approximately:

Saw logs	£35
Red logs	£28
Chipwood	£16
Paper pulp	£20

The saw logs tend to go for building timber,

the lower grade 'red logs' for pallets and fencing and the small roundwood for pulp and chip board. The brash, or small branches, are sold locally and chipped to surface playgrounds and horse arenas or for woodburning stoves. Future markets may include power stations and producing oil for generator engines. Some brash is left on site to help the forest vehicles travel about. Brash also offers shelter for succeeding new saplings.

After felling the areas are restocked. Natural regeneration is quite good and in some places the trees just need to be re-spaced to a density of 2,500 per hectare (roughly 2 m apart) when they reach 2 m in height. Most areas, however, need to be replanted and approximately 2 million seedlings are put in each year, almost all by hand.

Nature conservation

Within the forest there are ten SSSI's covering 7,500 hectares (4 per cent of the total SSSI area in England). Many of these are border mires with species such as *Sphagnum*, bog asphodel, long-leaved sundew and bog rosemary. Some 20 per cent of Kielder is always treeless and is left so, in order to promote wildlife.

Management specifically for nature conservation objectives includes the felling of trees prematurely on some of the mires which were planted up in the past, the planting of broadleaves which, in the twenty-first century will compose 10 per cent of the forest and the use of native tree species where appropriate. No income is obtained from the broadleaved trees so any increase in their area consequently results in a reduction in the productive land.

Management for specific species is not the norm, but Kielder is one of the places where red squirrels are still found. Since Sitka spruce has rather small seeds, there is now a policy of

ensuring that 17 per cent of the conifers restocked are other species (such as Scots pine and Norway spruce) to provide adequate food for the squirrels.

Visitors and public relations

In recent years there has been an increased emphasis on ensuring that local people are better informed about what is going on in the forest. This includes telling immediate neighbours about future felling plans. Access routes attempt to avoid sensitive areas and villages but extraction roads put in when the forest was planted are not necessarily in the best places today so there is continually a need to reassess the best routes.

Kielder also actively encourages visitors and there is free public access to all areas owned by Forest Enterprise except when work is being carried out with machinery. Rangers help to look after the public and organise walks, open days and other events. There are cycle routes and horse riding routes and the long-distance foot path, the Pennine Way, passes through. Most of the day visitors stay close to the main centres so the pressure is minimal in the more outlying areas. Forest Drive is a 19 km scenic drive through the middle of Kielder which is open between Easter and October.

One funding problem for Kielder is that the income from timber sales goes to central government. Money cannot be carried over from one year to the next so it is difficult to undertake big capital projects such as building toilet blocks or visitor centres.

INSHRIACH, CAIRNGORMS

The Cairngorms National Nature Reserve (NNR) totals 25,000 hectares in all and

Species Box 4.2: Scots pine

Scots pine is the large coniferous tree native to the north of Britain and Scotland. Many of those found in southern Britain derive from European stock whereas those native of Scotland belong to a separate variety, *scotica*. Scots pine can survive in a wide range of situations from very humid to surprisingly arid. It is regularly planted on sand dunes to help stabilise the soil. The trees grow to about 30 m tall and are usually simple and fairly straight. Older trees may develop flatter crowns. The bark is reddish in colour, when young it can be rather thin and papery, and as the trees get older the bark becomes thicker and plate-like. The needles arise in pairs and are dark green and 50–100 mm long with pointed tips. In usual conditions they survive for about four years. The male flowers are found in clusters and are pale in colour. The female flowers are dark pink and are found towards the top of the tree. The cones are quite small (about 50 mm long), and start roughly oval to egg-shaped, opening when ripe to release the seeds. Trees start bearing cones at about 16 years of age and produce maximum crops when 60–100 years old. The timber is widely used commercially where it can be called a variety of names. It is strong and light and is used for construction work and furniture. In the

past it was used extensively for pitprops, railway sleepers and telegraph poles. In commercial plantations the trees are felled at roughly 70 years of age.

Source: White (1995)

belongs to a variety of different owners. The forests of Inshriach and Invereshie (3,084 hectares) are owned by Scottish Natural Heritage (SNH), Upper Glen Avon by the RSPB and Mar Lodge, a fairly recent acquisition by the National Trust for Scotland.

The Cairngorms are the largest area of mountainous land over 900 m in Britain. The NNR includes a range of habitats from the open tops of the mountains to extensive woodlands. The native woodland here is largely dominated by Scots pine (Species Box 4.2).

After the last Ice Age much of Scotland was covered with woodland. This woodland consisted largely of broadleaved species, with conifers being dominant only on the poorer soils, particularly in the Central Highlands. It has been suggested that almost half this woodland disappeared between the Mesolithic period and the 1st century AD, due to climate change (wetter and windier weather which led to the formation of blanket bog) and human clearance. The first known published reference to the 'Caledonian Forest' was by Pliny in 77 AD and probably refers to the land north of the Firth of Forth, an area which was then occupied by the tribe known to the Romans as the

'Caledonii'. Deforestation continued and by the end of the Middle Ages it has been estimated that only 4 per cent of the land area of Scotland was wooded, and much of this was confined to the more inaccessible parts of the Highlands.

Today less than 1 per cent of the native pine forest remains (RSPB, 1993) and Figure 4.4 illustrates the former and current extent. Table 4.1 shows the reasons for losses in more recent times. Those woodlands that remain tend to be small remnants in the most inaccessible areas (Plate 7). Recent work has centred on protecting these remnants and increasing the amount of native woodland. RSPB (1993) highlights the plight of the native pine forests, outlines current initiatives and suggests methods for managing and increasing the area of native pine.

Until the eighteenth century the remaining native pine forests at the foothills of the Cairngorm Mountains (of which Inshriach is

Figure 4.4: The area of Caledonian pine woods today. The current area is shown in black, the former boundary depicted by a dotted line. From RSBP (1993).

one) were exploited mostly for local use as imported timber from Scandinavia and the Baltic was readily available throughout Scotland at competitive prices. During periods of war this trade was disrupted and exploitation of some of the more remote forests became viable.

Inshriach Forest, which then formed part of the Mackintosh estate, underwent considerable felling during the Napoleonic wars, between 1800 and 1850. The timber was extracted by horses as far as the River Feshie in the valley bottom and then floated out through a series of specifically constructed lochs and sluices. There was further felling during the 'Great War' (1914–18) and also during and immediately after the Second World War. The boundary of this last felling is still very obvious on the upper slopes today.

Much of the lower slopes were subsequently planted, mainly with Scots pine, by the estate and later by the Forestry Commission. Thus the upper areas have not been planted and the highest parts probably never felled. There has probably been very little domestic grazing in the area except prior to the nineteenth-century felling when the valley bottom was known as the 'thieves route' and was used by drovers for their cattle.

Today Inshriach can be divided into three main zones. On the lower slopes are the commercial forestry plantations consisting of Scots pine, which are deer fenced and managed by Forest Enterprise. The highest part of the reserve has a dense covering of Scots pine. This is the oldest woodland, is probably natural in origin and has a high density of trees. Between these two zones is an extensive area that consists of moorland and open woodland. This is the part that was felled in the nineteenth century, was not subsequently replanted, and is where natural regeneration of the native woodland is now being encouraged. The tree

Table 4.1: The loss of native pinewood in Scotland between 1957 and 1987

Causes	Area lost (hectares)
Underplanted with local origin Scots pine	443
Underplanted with other species or origins	1 984
Felling	1 308
Windthrow	42
Fire	110
Total Loss	3 887
Area of native pinewoods in 1957	15 900
Percentage loss	24

Source: RSPB (1993)

densities and ages are very variable. The ground flora consists of heather, cowberry (often referred to as cranberry in Scotland) and bilberry (blaeberry) with occasional species such as creeping ladies tresses and twinflower (the only site in the Cairngorms for this species). Birds include capercaillie, crested tit and Scottish crossbill (Britain's only endemic bird).

Until recently it was thought that the native Caledonian pine woodland was very open, with the trees widely spaced. In many places, however, the good timber trees were selectively removed leaving the poor ones behind and this encouraged the development of open woodland. In the few areas that have never been cut, the tree density is higher than that of a commercial plantation with up to 50 per cent of the trees dead but still standing (RSPB, 1993).

The current management aims for Inshriach are to encourage the regeneration of Scots pine and to safeguard the existing species of interest found in the reserve. This is to be achieved where possible by minimum intervention. No ground preparation is carried out and no planting takes place. Restocking with Scots pine is achieved by natural regeneration. This occurs throughout the open area as the seed

disperses well by the wind. The genetic origin of the regeneration is probably largely local and that from the trees on the higher slopes is considered native. The commercial plantations were probably planted using seedlings from Abernethy, which was one of the earliest pine nurseries. Most of the seed used at Abernethy was of local origin, thus the planted trees at Inshriach are probably local, if not derived from the site.

One of the biggest problems in most native pinewoods is the heavy browsing of young pines by deer and Inshriach is no exception. On many sites this can result in complete absence of young trees. Red deer damage to Scots pine is especially high in periods of heavy snow, in June and July when the new growth is soft and palatable and in October during the rutting season when the stags can thrash young saplings in a display of aggression.

In the mid-1970s when SNH purchased the land and then acquired the shooting rights at Inshriach some heavy culling of deer was carried out. Forty stag and forty hinds were shot annually which substantially reduced the numbers. A reduced cull of only ten stags and ten hinds and calves is now required to keep the population in equilibrium. Concern was expressed that when the numbers were reduced deer would just flood in from the surrounding area but this does not seem to have happened. One of the reasons may be the strong territorial nature of the hinds and their reluctance to change their territories. Following the initial culls a substantial amount of pine regeneration was able to grow (Plate 4.6) and much of this is now about 20 years old.

There is still some browsing by deer but it is at a level that is acceptable in terms of sustaining the forest resource. The deer fences are now quite 'porous' and there are more deer in the plantations now, especially during the winter months when shelter is important. The deer forage out from these areas so that the regeneration is more likely to be heavily browsed if it is close to the plantations. Thus that part with the best chance of surviving is furthest from the areas where the deer hide and this can be clearly seen on the reserve.

Pine is not the only tree species found in Caledonian pine forests. It is thought that about 30 per cent of the forest should be broadleaves. These include alder and willow in the damper areas and especially along the river-

sides, birch and rowan in among the pines, and dwarf birch close to the tree line. Juniper is another component of the forests. At Inshriach juniper regenerates quite well but birch and the other broadleaves are at a much lower density than would be expected. They grow well until they get above the height of the heather and are then browsed. The browsing is probably largely by roe deer, which are much more difficult to control than the red deer.

Monitoring is carried out on the reserve using line transects to record the tree regeneration. The British Trust for Ornithology are also monitoring the changes in bird populations as the forest develops in the areas which are quite open at the moment.

COED Y RHYGEN, GWYNEDD

On the west of Britain the high rainfall and unpolluted air result in woodland that abounds with bryophytes (Species Box 4.3). The Merioneth oak woods are fine examples

Plate 4.6: Regeneration of Scots pine, Inshriach

. . . .

Species Box 4.3: *Thuidium tamariscinum*

Thuidium is a very common and distinctive moss. The general shape of the branches is very much like the frond of a fern, each being three times pinnate and very regular. The branches can stand quite upright giving the plant the appearance of being a deep but loose mat. The small, scale-like leaves and the branches are rather sparse towards the base of the plant and the stem blackish-green. Towards the apex the plant is much greener in colour. *Thuidium* reproduces sexually by a large capsule full of spores on a red stalk but this is rarely seen. This species of moss is very common in woods and hedges especially on heavy soils. It may also occur in more open conditions in the north and west.

Source: Watson (1981)

. . . .

of native woodlands that result from these climatic conditions. Plate 8 shows one such woodland. Coed y Rhygen National Nature Reserve is a 35-hectare woodland, of sessile oak and downy birch, which is categorised as W17 using the National Vegetation classification. Three of the four sub-communities are present and there is a small area with hazel. Apart from some silvicultural management on a small part of the reserve some years ago, and some farm grazing, the wood is generally con-sidered to be one of the least disturbed in Wales. The majority of the wood is owned and managed by Countryside Council for Wales for its nature conservation interest alone. There is no public access and no commercial timber crop is taken but the wood is grazed by domestic sheep. It is generally considered that, at present, the condition and structure of the woodland and the bryophyte communities are good with just a few small areas where, for example there is good natural regeneration but few older trees. Coed y Rhygen has a management plan written in the Countryside Management System style and thus important attributes are identified and upper and lower acceptable limits are set for them. The aim of the management is to keep within these limits. Monitoring projects are designed to ensure that the management is achieving its desired aims. Some examples are given below.

The management aims to maintain the woodland structure in a favourable condition, and ensure that the bryophyte communities are maintained. As a general principle the management aim is that of limited intervention, as described below.

1 Keeping the woodland area at 35 hectares, not allowing it to extend into surrounding, non-woodland areas, of high conservation value and not decreasing in size. This is monitored using aerial photographs taken every five years.
2 Maintaining the tree canopy to have a cover of between 70 and 90 per cent. Monitored by aerial photographs.
3 Allowing a gap creation rate of between 0.25 and 1.5 per cent of the woodland area per year. Monitored by aerial photographs.
4 The composition of the canopy, sub-canopy and shrub layer should continue to represent the range of native species found locally, with at least 60 per cent oak in the canopy. Monitored by aerial photographs.
5 The natural regeneration of native trees in the canopy gaps over a 20-year period should be at least 3 per 0.1 hectare of gap. These should reach a minimum height of 3 m and be of species native to the woodland.
6 The amount of dead wood (as fallen trees, branches on the ground and in trees and standing dead trees) should be not less than 30 cubic metres per hectare. This is moni-

tored by surveying the volume of dead wood every ten years.
7 Ensuring the field layer does not suppress the ground layer (limits to be set, and monitored, after baseline survey).
8 Ensure that there is no loss of bryophyte species. Monitored by permanent quadrats.
9 Ensure that the populations of nationally scarce and indicator species are maintained at, at least, baseline levels. Monitored by permanant quadrats after notable species have been located.

Survey work and archival records, e.g. aerial photographs, are used to establish the upper and lower limits given in some of the categories above.

As the woodland is generally in good condition, much of the work needed is monitoring to make sure that it remains within specified limits. There are two main areas where something more active needs to be done. The first is with regard to rhododendron. This species is a big problem in the woodlands of the surrounding area and Coed y Rhygen is no exception. Vigilance is needed in order to ensure that any colonising rhododendron is removed and does not become established.

The second problem is the issue of grazing. The woods are grazed by domestic sheep and, in the past, by feral goats. While occasional visits by the goats are not a problem, in the past there was a herd locally which it is considered should not be allowed to build up again because of the potential damage to all the local woods. It is much easier to control the numbers of domestic sheep but difficult to set ideal limits on numbers. High grazing levels reduce the field layer and create more open conditions, which is better for the bryophytes. Lack of grazing results in substantial growth of the field layer which is detrimental. One sheep per acre per year seems to be the ideal level for bryophytes. Unfortunately, at this grazing pressure natural regeneration of the trees does not take place; one sheep per two acres seems to be a better level to achieve this aim.

The solution to this conflict is to alternate the grazing pressure over a long period of time. Thus the higher stocking rate will be used most of the time but every twenty years (in the majority of the area) the grazing pressure will be relaxed for long enough to allow tree regeneration but continue to restrict the development of a field layer.

THE NATIONAL FOREST, THE MIDLANDS

The National Forest is an ambitious plan which was conceived by the Countryside Commission in 1987. The intention is to turn 50,000 hectares (or 200 square miles) of countryside to the north east of Birmingham into a new forest. Like the Royal Forests of the Middle Ages the National Forest is planned not just to be woodland. It is anticipated that approximately one-third will be covered with trees, another third agricultural land and the remaining third residential areas, infrastructure and other types of open land. In 1987 the tree cover of this area was just 6 per cent with agricultural land predominating, there were also substantial areas of derelict land which could easily be planted up with trees. The extent of the new Forest includes parts of Staffordshire, Derbyshire and Leicestershire with the towns of Burton upon Trent, Swadlincote, Ashby-de-la-Zouche and Coalville as well as a range of smaller, mostly agricultural settlements. The aim is for 30 million new trees to be planted as a mixture of native broadleaves and conifers with 70 per cent of the planting to be achieved in the next ten years (see Plate 4.7).

The aim of the National Forest is that of a sustainable multipurpose area. More specifically, Bell (1993) lists the objectives as:

Plate 4.7: Young mixed deciduous trees protected by Tuley tubes, one year after planting.

1 a functional and working forest (though under multiple ownership and management);

2 an environmentally sensitive area, blending the trees into the historical, ecological and cultural character of the area;

3 geographically diverse, with a mixture of land uses, varying in intensity;

4 a sustainable forest, with appropriate and related development being welcomed.

This should be achieved by the following means:

1 the planting and management should be done by choice, not compulsion;

2 the strategy should be achieved in a politically acceptable and cost-effective manner involving the private and voluntary sectors;

3 full use should be made of all existing measures to encourage planting and other forest-related activities. New measures should be introduced if there is a short fall in these mechanisms.

The Forest should be able to produce a commercial timber crop and also promote the development of farm diversification and rural industries. One of the reasons that this particular area of Britain was chosen for the location of the Forest was because it is a region of economic and social hardship, following the closures of the mines. Therefore new business opportunities and jobs in the wake of the Forest would be welcomed.

Another main function of the Forest is to provide a recreational resource for both local people and visitors. This is seen as both informal recreation and organised activities. Land-owners are not compelled to provide public access but there is a good right of way network through the area. It is anticipated that the Forest will be accessible to visitors through the public transport system and that the towns and villages should also become an integral part of the Forest.

Initially, after the chosen site for the Forest was announced in October 1987, and the first trees planted, the Countryside Commission remained as the lead organisation. The National Forest Development Team was set up with the aim of promoting public interest and stimulating tree planting. The Team also prepared a Forest strategy and a business plan. Then in 1995, with substantial government backing, the National Forest Company was set up. The function of the Company is to achieve the task of developing the National Forest. The way forward was seen to be in the form of partnerships between local authorities, landowners, companies and the local people.

In a more tangible way, the Company is able to help financially with the costs related to tree planting and changing the focus of activities to those related to woodlands. This is done in the following ways:

1 The Tender Scheme. This system encourages landowners to convert their land to woodland and wood-based activities by providing funding. Some 76 per cent of the tree planting in the first two years of the Company has been carried out through the tender system. However, the scheme has a limited life. An example of the type of projects funded by the scheme is Beehive Farm. Situated near Swadlincote, Beehive Farm used to be a mixed dairy farm which, in the current economic climate, was ceasing to be financially viable. With the help of the Tender Scheme the farm's owners have now planted 26,000 trees which are aimed to produce a commercial crop in the future. They have also created two fishing lakes,

established a caravan site and developed a craft centre and tea room.

2 Grant Schemes. The Company is able to assist with projects involving woodland creation. This may be by buying land or assisting others to buy land.

3 Major Forest-related bids. Applications for funding on a large scale, for example through the National Lottery or the European Union, is time-consuming. The Company is able to apply in its own right and to help others apply. In 1996–7 £4.5 million was won, to be spent over a 4-year period.

4 Developing markets. Help is given with developing timber-based markets and also leisure and recreation facilities.

5 The public can have trees planted for them, for commemorative or celebratory purposes for a fee of £10 a tree and this generates money for planting.

The results so far have been tangible. The 2 millionth tree was planted in 1998 and the fundraising in 1996–97 was £35,460 with a further £80,000 given in gifts of kind. The recreational side is flourishing, with various leaflets being produced outlining walks and places to visit. Horse trails and cycle trails have been set up and signage and interpretative boards erected. Substantial funding has been secured from the Millennium Commission for the building of a National Forest Discovery Centre, a leisure and education centre to present information on the importance of trees and woods.

5

PRACTICAL WORK

•

By far the best way of understanding the ecology of a habitat is to investigate it for yourself. In order to gain maximum benefit from ecological studies it is important to plan in advance. Part of the planning is to be sure of the questions to be asked or problems to be solved; the most elegant research tends to ask simple questions or look at simple problems. Prior planning also includes identifying the methods for data analysis (see below). Equipment should be checked and the experimenter should be fully competent in its use before gathering data. A consistent approach, using the same techniques and only varying the factor to be analysed, is usually best. Often a small pilot study will help the main investigation to run smoothly and will allow the methods to be refined.

It is important that permission is received from the owner(s) of land being used for the project, both to gain access and to carry out particular types of experiment. For example, it is against the law (Wildlife and Countryside Act, 1981) for any unauthorised person (i.e. anyone who is not the owner, occupier or has been authorised by the owner or occupier of the land concerned) to uproot any wild plant. Other organisms are even more strictly protected; almost all birds and some other animals, together with several plant species, are fully protected under the law (see Jones, 1991, for further details). Care should be taken to comply with the law, to ensure that you do not damage or unduly disturb any plant or animal. The research project should be designed and implemented so as to leave the habitat as it was found.

Woodland habitats provide a wealth of topics, which may be investigated throughout the year with relatively little specialised equipment. The following projects may be investigated with relative ease, require little in the way of equipment and utilise common woodland habitats or species. They are designed to ask simple questions and provide data to analyse with simple tests. As in most research, the results obtained may stimulate further work on supplementary questions.

EXPERIMENTAL DESIGN

Comparisons between situations in which only one factor varies are the easiest to interpret. For example, comparisons between several sites differing in size, but similar in all other respects will allow an examination of the influence of size of site on whatever is being recorded. The experimental design must be considered in some detail. There are two major types of experimental design: observational and manipulative. The majority of projects described below are observational. Here, a variable (the behaviour of animals, the percentage cover of different plant species, the numbers of animal of a particular species, etc.)

is recorded under different circumstances (different sites, weather conditions, numbers of animals in a group). Analysis is often a matter of finding whether the variable measured differs in two or more circumstances (e.g. sites, times of day, etc.). If two variables (e.g. number of species and temperature) are measured, you may wish to examine the relationship (called correlation) between the two. Two variables may have a positive (as one increases, so does the other), negative (as one increases, the other decreases) or no relationship. However, with observational experiments you will not be able to say definitively that a change in one variable causes the change in the other. This is because an observed increase in one variable (e.g. the number of a species) may be correlated with a measured rise in another factor (e.g. temperature) but actually be due to changes in a third, unmeasured variable (e.g. the abundance of its prey). In manipulative experiments, on the other hand, the experiment is designed such that one variable is altered by the experimenter (pH, temperature, fertiliser concentration) allowing a much greater emphasis on cause and effect when it comes to analysing the data. However, since these are often conducted in fairly artificial conditions, they may have less relevance to real-world situations than may be achieved with observational experiments.

It is important to take several replicate samples, since data gathered from only a single sample may not be representative of the situation in general. In order to reduce bias, sampling should be systematic or random: instructions on sampling are given in each practical. It is essential that data gathered are recorded correctly to avoid later problems in analysis and interpretation. In order to help you collect data in an appropriate way, sample data recording sheets have been included where

appropriate. Record your data in a hard-backed book rather than loose sheets of paper to reduce the risk of loss. Where data are to be analysed using a computer, it is useful to transcribe the data onto an appropriate spreadsheet or other data file as soon as possible, and to check the transcription carefully to identify and rectify any mistakes. Techniques for the analysis of such data are beyond the scope of this book although methods of analysis are suggested in the project descriptions, and several texts which provide further details are listed in the further reading section.

HEALTH AND SAFETY

Health and safety in field (and laboratory work) should be paramount. Do not engage in behaviour or activities that could harm yourself or others, and assess the risks and health and safety issues which are likely to be involved and protect against them. General safety issues include: wearing appropriate clothing for the time of year and terrain; using safe equipment (e.g. plastic tubes rather than glass); not working far from help (ideally work in groups of two or more) and informing someone responsible of the details of planned field work in advance (including location and duration) and 'signing off' with that person on return.

Soil and water contain organisms and compounds, which are hazardous to health. Gloves should be worn for all fieldwork involving these materials, and when handling spiny, or otherwise hazardous plants. Tetanus is a potential hazard for anyone working out of doors, especially those in contact with soil. Spores of the tetanus bacillus live in soil, and minor scratches (e.g. from bramble thorns) could provide a point of infection.

Immunisation is the only safe protection and, being readily available, should be kept up to date. Weil's disease (or Leptospirosis), is caused by a bacteria carried by rodents (especially rats). Urine from infected animals contaminates freshwater and associated damp habitats such as river, stream and canal banks, and is more common in stagnant conditions and during warmer months. Infection is usually via cuts or grazes, or through the nose, mouth or eye membranes, and precautions should be taken to avoid contact between these areas and potentially infected water. Cover cuts and grazes with waterproof dressings, use appropriate waterproof clothing including strong gloves and footwear, and avoid eating, drinking or smoking near possible sources of infection. Lyme's disease is another potential hazard for fieldworkers. This is transmitted by female ticks, which can be found in the uplands and some woodlands. They are especially likely to bite from early spring to late summer. To avoid the disease, prevent ticks from biting by wearing appropriate clothing (e.g. long trousers). When returning from the field check for the presence of ticks (light-coloured clothing helps) and remove any found as soon as possible, if bitten (twist slowly in an anticlockwise direction without pulling and seek medical help if mouth-parts remain within the skin).

In addition, if working in more urban or disturbed woodlands watch out for obvious hazards such as sharp pieces of metal, plastic and broken glass and avoid traffic hazards when working close to roads.

The safety issues indicated here are not comprehensive. Before any fieldwork is undertaken you are advised to consult appropriate publications such as that produced by the Institute of Biology (Nichols, 1990).

STANDING VOLUME OF TIMBER IN COMMERCIAL FORESTS

Commercial forests provide unusual circumstances for biological projects involving fieldwork. Because the trees in a block are usually all the same species and were planted at the same time, the individuals sampled are known to be from a relatively homogeneous population. This enables comparisons to be made between different species in the same area or the same species in different situations. By talking to a forester working on the site, or the site manager, two areas can be found where the same tree species was planted in the same year and their growth in different compartments can be measured.

In commercial situations the volume of timber in each standing tree is calculated by estimating the tariff number for the tree species in a particular situation and then using the volume tables. Full details and the relevant tables can be found in Hamilton (1985). If this book is not available, estimates of volume can be calculated by the following method.

For conifers, the useable wood is just the trunk of the tree and this can be assumed to be a cone in shape. It is therefore necessary to measure the height of the tree, which can be done using a simple clinometer (see Figure 5.1). The diameter of the tree needs to be established at ground level, above any root spurs. The diameter is calculated from the circumference of the tree (easily measured) as:

$$d = \frac{C}{\Pi}$$

where d is the diameter and C is the circumference. The volume of the wood in the trunk of the tree is then calculated as:

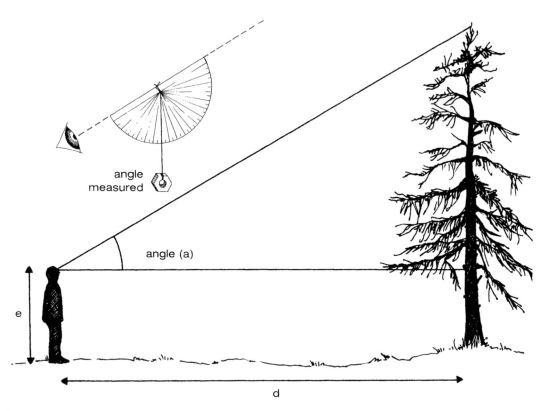

Figure 5.1: A simple clinometer. This can be made from a protractor which has a small hole drilled in the centre, through which is threaded a piece of fishing line with a weight on the end. Stand on the same level as the tree and about 10 m away (measure this distance: d). With the flat side of the protractor uppermost and allowing the weight to hang freely, sight along the protractor to the top of the tree and measure the angle along which the weighted line hangs. The angle to the top of the tree (a) is calculated as 90 minus this angle. Measure the distance from your eye level to the ground (e). The height of the tree can be calculated using the following formula (making sure that all your measurements are in metres):

Height of the tree = (the distance from the tree × the tangent of the angle) + your eye height

$$\frac{\Pi d^2 h}{12}$$

where d is the diameter and h the height.

It is necessary to measure at least ten trees from each plot (ensuring the same number from each). Table 5.1 is a suitable recording sheet for this project. The mean and standard deviation of the volume can be calculated and the samples analysed using t-tests. By counting the number of trees in a measured area a total volume of timber per hectare can be calculated and an estimate of the value made if required (ask the forester, or contact Forest Enterprise, for current timber prices). Any differences found can be related to factors such as aspect, soil type (pH and approximate soil type are easily determined), density of trees, amount of light reaching the trees and any differences in management.

Table 5.1: Recording form for the volume of timber in commercial forests

Tree	Location	Species	Height	Circumference	Diameter	Volume
No. 1						
No. 2						
No. 3						
No. 4						
Etc.						

If it is not possible to find plots of the same tree species planted in the same year, then comparisons of different species can be made or the same species at the edge and in the middle of blocks (i.e. subject to different light conditions).

THE DECOMPOSITION RATE OF LEAVES IN WOODLAND

The rate at which leaves decompose varies according to a range of factors, for example some tree species are more palatable than others and tend to decompose quicker. The agents of decomposition range from microscopic bacteria to large arthropods; some physically breaking down the leaves and others acting chemically to reduce the leaf fragments into their components. This project looks at the relative effectiveness of different-sized decomposers in breaking down leaves.

Small bags are made to contain the leaves. Net curtain material is quite suitable but other types of netting can also be used. The stronger the material the better and nylon is ideal. Two different types need to be used, one with very fine holes of roughly 0.5 mm or less, the other with larger holes, up to 5 mm diameter. The material should be made into bags of roughly similar size, about 50 mm square when complete. When sewing up the bags ensure that the stitching is not looser than the hole size of the bags, it is best to use a sewing machine if possible. Into each bag place some leaves of a known tree species (preferably one found at the site being used), approximately the same amount in each bag and then sew the bags shut. You will need at least ten bags of each mesh size (the more the better). When the leaves are sealed in the bags they need to be dried in an oven until they reach a standard weight. The bags must be numbered (e.g. using a marker pen) so that the original weights are known accurately. Then all the bags can be put out into a woodland, preferably where they can be easily found again. Ideally they should be partially buried into the leaf litter, not just placed on top. They can be left for a predetermined time according to the length of the project. Three months is probably the minimum time to leave them but in the spring or autumn it might be possible to obtain results in a shorter period. When the bags are collected they must be dried again until they remain at a constant weight and the amount recorded. It may be wise to shake the bags before weighing to remove any soil sticking to the outside and also to eliminate any soil which has 'fallen' into the bag (especially those with a large mesh size). The weight loss (i.e. the original weight minus the final weight) is an indication of the amount of decomposition that has taken place. The bags with the smaller mesh size will allow only the smaller organisms to enter. You may want to have another set of bags with filter paper in to act as a control. Table 5.2 illustrates a suggested recording sheet for this project.

There is plenty of scope for expanding this project, for example:

1 replicate bags with different tree species in to see which are decomposed quickest;
2 as above but comparing deciduous and coniferous species;
3 replicate bags in deciduous and coniferous woods to look at relative rates of decomposition;
4 using bags with a range of mesh sizes;
5 using more bags but collecting some in at intervals over a year to look at the rate of decomposition over time;
6 combining with invertebrate sampling to gain an idea of which decomposers may be present on the site.

The mean weight loss can be calculated for the different categories of bags and these can be compared using t-tests. Interpretation of the results may include consideration of the type of animals involved in the decomposition at the particular site and links can be made to the depth and type of leaf litter found in the woodland.

TREE GROWTH

Each winter deciduous trees cease growing, and at the end of each branch a terminal bud forms, which remains dormant. In the spring the twigs start growing and elongate from this bud. When the growth starts the scales covering the bud drop off and leave a set of scars. These scars can be distinguished from where a leaf has fallen because they make a ring around the twig. They are often known as terminal bud scars or girdle scars. By looking carefully at a twig and working backwards from the tip it is usually possible to distinguish previous years' terminal bud scars. The length of twig between them is the amount the twig has grown in a particular year (Figure 5.2). This technique can be used to compare growth between trees or even between different sides of the same tree (for example, if it is growing at the edge of a clearing). In woods where there is abundant regeneration, the growth of saplings can be measured and the speed of growth related to the amount of light they receive.

Locate a selection of saplings in a range of situations within the wood. Ideally some should be in quite dense shade and others in the open in a clearing. It is important that they are all the same species (although the project can be extended to compare two species). For

Table 5.2: Recording form for the decomposition of leaves in woodland

Bag no.	Mesh size	Leaf species	Location in wood	Time left in wood	Original dry weight	Final dry weight	Weight loss
1							
2							
3							
4							
Etc.							

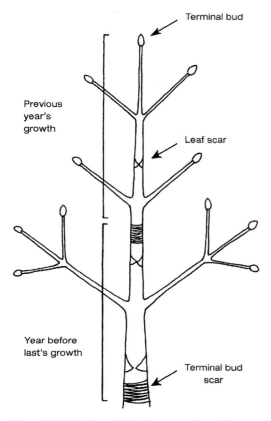

Figure 5.2: Using terminal bud scars to determine growth on saplings.

For each sapling the amount of light it receives needs to be found. This could be done just using a subjective scale of, for example, 1 to 10 but it is better to measure it more accurately. Ideally the light levels at different wavelengths should be measured but in reality the equipment is not usually readily available to do this. A camera light meter is probably the easiest method though this may give readings which are higher than the amount of light at the wavelengths most useful for plants. See Table 5.3 for a suitable recording sheet for this project.

The mean amount each sapling has grown in the last three years can be correlated with the amount of light it receives, use a rank correlation method if you are estimating the light using a subjective scale. Alternatively, you can analyse each year's growth separately. If the wood is suitable and the equipment to measure the amount of light is not available, it may be possible to compare growth in a large clearing or at the edge of a wood with that in an area of dense canopy. In this instance the Mann-Whitney U test can be used.

VISITORS' EXPERIENCES AND PERCEPTIONS OF WOODLANDS

There is an increasing trend for areas of woodland to be opened to the public. The types of site range from wooded corners of urban

each sapling measure the distance between the terminal bud scars on the leading (i.e. longest) shoot. If possible, measure several years (e.g. three) not just the last or current season, ideally the same number of years for all the saplings.

Table 5.3: Recording form for sapling growth based on terminal bud scars

Sapling no.	Species	Distance between terminal bud scars			Amount of light
		Current year	Previous year	2 years previously	
1					
2					
3					

parks to large blocks of commercial plantations. In addition, many country parks contain woodland. While the numbers of visitors to some of these parks can be extremely large, people have varied expectations and experiences. Although visitors might like to drive through areas with dense tree cover, they often tend to congregate in more open patches. If woodland areas are being planted with public access in mind it is important that aspects of public perception are taken into account. This is also an important point to consider when looking at existing woods with public access or where an increase in visitor numbers is desired.

To carry out this project it is necessary to find a site which has public access and a range of habitats, preferably including woodland of differing density and structure, and some open areas. It does not need to be entirely woodland; sites which are partly heathland, grassland or similar are also suitable.

A simple questionnaire can be drawn up to obtain information from visitors to the site. This should not be too long otherwise people will not feel inclined to stay and finish it. The questions should almost all be 'closed' (i.e. the responses should be yes or no or fit into a scale of, for example, 1 to 5). This makes the analysis much easier. Open questions are best avoided. Examples of closed questions are:

On your visit to Anyplace Country Park today have you visited 'Devil's Coppice'?
  Yes No Don't know
Or

How often do you visit 'Devil's Coppice'?

1 Never
2 Very occasionally
3 Occasionally
4 Often
5 Frequently

An open question might ask 'What do you like best about Devil's Coppice?'

To start the questionnaire it is useful to have a few general questions about how far visitors have travelled and how frequently they come. Local or frequent visitors may behave in different ways to first-time visitors. It is also helpful to record details such as the weather conditions. For the major part of the questionnaire it is important to be clear about the objectives and hypotheses the study is testing. Suitable issues might be that visitors show no particular preferences for different parts of the site, or that regular visitors visit the same areas of the site as occasional visitors.

If there are different types of paths on the site it might be possible to compare the preference for people for a narrow path in amongst tall trees to one which is wider with graded edges. With all projects it is important to obtain permission from the owners/managers of the woodland but with this project it is often helpful to involve them in the formulation of the questions. Managers may have issues of importance to them, which an extra question might answer. This is likely to make them more helpful over granting permission, etc. A talk to the manager may also help to focus the project on a particular issue, which can help with the management or policy on the site. Do not forget to send them a summary of the results when the project is completed.

When approaching members of the public to request help in completing the questionnaire always be polite and explain what you are doing and why (e.g. project for college). It is a good idea to have some form of identification (NUS card, or similar) in case of query and some sites will now provide you with cards themselves. When approaching people be obvious and do not stand behind a tree and ambush them. If they decline to help, then accept the rejection gracefully and find someone else.

You should aim for at least 50 completed questionnaires to get a reasonable sample size. Analysis will depend on the questions asked but if they are mostly yes/no answers or a range of 1–5, a good start is to tabulate them (an example recording form is shown in Table 5.4). The questions can be analysed individually (what percentage of people never visit 'Devils Coppice'?) or in pairs (what percentage of frequent visitors to the site never visit 'Devils Coppice' in comparison to infrequent visitors?). Suitable statistics for analysis may include Mann-Whitney U, rank correlation and Chi squared. Preliminary visual assessment of the results can be done using pie charts and histograms.

When interpreting the results, points to consider might include how the site could be altered to improve visitor experience or how it might be possible to 'encourage' visitors to less sensitive areas and away from rare species or vulnerable habitats.

ASSESSING THE ABUNDANCE AND DISTRIBUTION OF SMALL MAMMALS IN WOODLAND

A range of species of small mammals can be found in woodlands. While wood mice are usually the most abundant, shrews, voles and even dormice can be present. In a wood where there is a good shrub layer, some of these mammals use the above-ground area for foraging, as an extension of the woodland floor.

It is possible to use special live traps (longworth type) to catch small mammals but they do have some problems. The first is that they are very expensive to buy. To obtain enough for a survey is often beyond reach of many people (and institutions). The cheaper versions, made of plastic, are not robust enough and many animals can chew their way out. It is also essential that they are checked frequently to free any caught animals. A licence is needed from English Nature if it is anticipated that the traps might catch shrews. Most traps will not do so, unless set on a very sensitive setting, thus shrew numbers are usually underestimated. If the traps do catch shrews, they must be checked every few hours.

There are other methods of assessing the numbers of animals without using traps. Small mammals frequently use tunnels and narrow passages in the vegetation. Above ground they have to run along branches. A small tube placed on the ground or on a branch can be an attractive route for them. Square or round section plastic drain piping about 100 mm in width diameter (that from DIY shops is ideal) should be cut into sections about 0.3 m long. Inside the tubes, in the middle of one side, fix

Table 5.4: Recording form for questionnaire project

	Answers to questions					
Respondent	1	2	3	4	5	etc.
1						
2						
3						
4						

an inkpad (or a small piece of sponge soaked in ink). On either side of the inkpad fix a small piece of paper (see Figure 5.3). When the tube is put into position, with the ink and paper on the bottom, any small animal running through it, in either direction, will run through the inkpad and leave inky footprints on the paper. The disadvantage of this method over the trapping system is that it may not be possible to identify the species. However, an assessment of frequency of passages can be obtained giving an indication of the numbers of mammals. Books such as Brown *et al.* (1984) or Bang and Dahlstrom (1980) may help to identify the tracks as far as possible.

The tubes can be used to compare the numbers of mammals running about on the woodland floor with those above ground. At least ten tubes are needed for each habitat type. It may be helpful to position them in a rough grid so that they are easier to locate again but try to be flexible and place them in suitable places. The above-ground tubes should be fixed to low branches which the animals may run along so avoid any which are very thin or do not lead anywhere. They need to be firmly fixed in place so they do not fall off. Those on the ground are best put near to some cover rather than out in the open. They should be put on the top of the leaf litter but need to be settled so that they do not move about, but are where the animals can easily get into them.

Most small woodland mammals are nocturnal so it is important to leave the tubes out overnight. It may be necessary to experiment initially to perfect the technique and to find the optimal length of time to leave the tubes out in order to obtain some foot prints but not too many that it is impossible to see how many animals have passed through. In most situations one night is likely to be good enough. Once the system is working well, the tubes should be put out in the same positions for several nights in a row to obtain enough data for analysis. At the end of each night the paper can be collected and fresh paper put out. The number of animal tracks across each of the two pieces of paper in a tube can be added together to give the total number of animals passing through that tube in one night. Over a period of, say five nights, a mean number of animals passing through each tube can be calculated. Any differences between the numbers

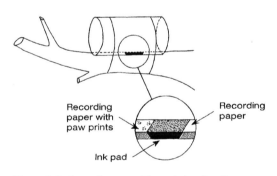

Recording paper with paw prints

Recording paper

Ink pad

Figure 5.3: A small mammal tunnel showing the position of inkpad and recording paper.

Table 5.5: Recording form for the use of tunnels by small mammals

Tube number	Habitat type	Night	Total number of tracks	Number of woodmice	etc.
1					
2					
3					
etc.					

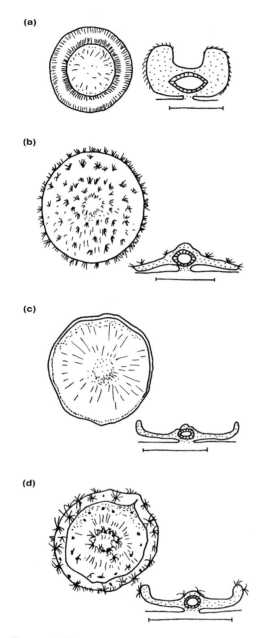

Figure 5.4: The identification characteristics of species of gall found on oak leaves, each viewed from above and in cross section. (a) *Neuroterus numismalis* (scale bar = 1 mm). (b) *Neuroterus quercusbaccarum* (scale bar 2 mm). (c) *Neuroterus albipes* (scale bar 2 mm). (d) *Neuroterus tricolor* (scale bar 2 mm). *Source*: Redfern and Askew (1992).

on ground tubes and above ground tubes can be tested using Mann-Whitney U tests.

A suitable recording form is shown in Table 5.5. If it is possible to identify different species, or groups of species (i.e. mice and shrews), then the analysis can look at these groups separately. If the woodland does not have a shrub layer which is developed enough to be able to use the tubes above ground, comparisons could be explored between different parts of the wood, for example, the edge and the centre, clearings and dense woodland or different types of woodland.

THE DISTRIBUTION OF GALLS ON OAK LEAVES

Spangle galls on the underside of oak leaves are very common and abundant in the late summer and autumn just before the leaves fall from the trees. They are made by tiny gall wasps with a complete life cycle showing alternation of generations. In the spring they form currant galls on the catkins of oak trees (which are hard to find). In the autumn the next generation causes spangle galls on the leaves which drop off after the leaves fall from the trees. The galls then over-winter in the leaf litter. In spring the adults emerge and lay eggs in the young, forming catkins. There are three widespread species which are quite easily identified with the aid of a hand lens. Oak is a very common tree species in woodland so it is possible to do this project almost anywhere. Variations can also be done using birch trees (which also have several different types of gall) or leaf miners in holly or hazel. Redfern and Askew (1992) provide good keys to galls separated by the species of plant they are found on. Full details of life cycles, predators, etc. are also given.

Collect 100 leaves, ten leaves from each of

Table 5.6: Recording form for the gall wasp project

Position of tree	Leaf	Species found	Number on leaf
1			
2			
3			
etc.			

ten oak trees, in each of two positions (e.g. edge of wood and middle of wood). Make sure the sample is equivalent from each tree. For example, collect either leaves from a variety of positions on the tree (different heights and aspects), or from all the same (low branches on the south side).

For each leaf in turn record the species of gall (use Figure 5.4) found on the underside and the distance each gall is from the leaf base (a suitable recording form is illustrated in Table 5.6). For each species found, calculate the mean number of galls per leaf. The numbers per leaf can be compared using *t*-tests or Mann-Whitney U tests. To see if there are significant differences between the distribution of galls of the same species on different tree types (or different trees) use the commonest species found.

This project can be expanded if collections of leaves are made from trees in other situations (in open conditions versus deep shade) or from different parts of the tree (south side versus north). If leaf miners are studied rather than galls, this aspect is worth studying because it is unlikely that more that one species will be encountered on each leaf.

GLOSSARY

•

Afforestation   Planting trees on unwooded land.

Ancient woodland   Woodland which derives from 1600 or earlier and has not been felled.

Ancient woodland indicators   Species which are likely to be only (or mostly) found in ancient woodlands.

Aspect societies   Communities of field layer plants which occur in the same area but at different times of the year.

Bolling   The trunk of a pollarded tree.

Brash   Small branches cut from trees. Can also be used as a verb meaning to cut small lower branches off plantation trees.

Broadleaved tree   One bearing broad leaves as opposed to needles.

Canopy   The highest layer of vegetation in a woodland, consisting of the tree tops, usually over 5 m in height.

Clear felling   Removing all the trees from an area.

Climax   The vegetation which results when an area is left to develop naturally dependent on the climate and soils.

Commercial forest   Woodland managed with economic reasons as the first priority.

Conifer   A cone-bearing tree.

Coniferous woodland   A wood of coniferous, or cone-bearing, trees

Coppice   A tree or shrub which has been cut at, or close to, ground level. A crop of branches results, these can then be harvested in future years. As a verb it means to cut a coppice.

Crown   The part of the tree consisting of the branches and foliage.

Deciduous   Trees which lose their leaves in the winter. Is also applied to woodland.

Diameter at breast height   The standard measurement of tree girth, taken 1.3 m above ground level.

Epiphyte   An organism like a moss or lichen which lives on the branches of a tree but is not parasitic.

Evergreen   A tree which does not shed all its leaves in the autumn months.

Field layer   The herbaceous plant layer of a woodland, usually between 0.1 m and 2 m above the ground.

Forest   See Royal Forest and Commercial forest.

Forestry   The management of woodland to produce timber commercially.

Ground layer   The vegetative part of the wood which is closest to the woodland floor, usually up to 0.1 m in height.

Hardwood   The timber of broadleaved trees (which may or may not be hard).

High forest   Woodland which is not managed as coppice or pollards. It may or may not be managed for timber.

Layer   The creation of a new shrub by laying an existing branch along the ground from where roots arise. This may happen naturally or may be a result of active management.

Maiden   A tree which has not been cut back.

Mast   Seed (e.g. beech nuts) which is produced as a heavy crop in some years.

Minimum intervention   Woodland management when almost no work is carried out. Minimum intervention can be due to a conscious decision or neglect.

Mycorrhizae   The association between a plant and a fungus living on or in its roots. The relationship is thought to be beneficial to both partners.

Native species   Those which are considered to be naturally occurring in the UK.

Natural regeneration   Young trees resulting from the germination of seeds on the woodland floor which occur as a consequence of natural seeding.

Natural woodland   Woods where there is no intervention in terms of management.

Needle   The leaf of a coniferous tree which is long and narrow in shape.

NNR   National Nature Reserve.

Non-intervention   The management of woodlands when nothing is done. In the UK most non-intervention woods are more accurately described as minimum intervention.

Nurse crop   A crop of trees which is grown to encourage or protect another species in its early years.

NVC   National Vegetation Classification.

Pannage   The practice of turning pigs out into woodlands during the autumn.

Park   An area enclosed for keeping deer. More recently, an area enclosed for aesthetic or amenity reasons. It may or may not have been formally landscaped.

Photosynthesis   The process by which plants convert water and carbon dioxide into sugars and oxygen in the presence of light.

Plantation   An area of trees planted for commercial reasons.

Pollard   A tree cut 2–4 m above ground level, producing a crop of branches which can be harvested in subsequent years. As a verb it means to cut a pollard.

Primary woodland   Woodland which has never been cleared.

Provenance   The region where a tree comes from or where the seed is collected from.

Pruning   Cutting live branches on timber trees to improve the wood quality and value.

Pulpwood   Timber which is for the production of wood pulp. It is usually small diameter branches in lengths of 2–3 m.

Recent woodland   Woodland which is not ancient (i.e. derives from more recently than 1600).

Ride   A track in a wood.

Roundwood   Timber which is of too small a diameter to be sawn in a mill.

Royal Forest   An area of land set aside by the king for hunting. It may include woodland and also open areas.

Saproxylic   Species which are dependent on dead or dying wood or other organisms associated with this habitat.

Saw logs   Timber which is to be sawn in a mill.

Scrub   An area of land dominated by shrubs, often mid-way between being open and covered with trees.

Secondary woodland   Woodland which has developed naturally on open land where once there were trees.

Semi-natural woodland   Woodland that contains native tree species and maintains itself by natural processes.

Shrub layer   The woodland layer which is dominated by shrubs. Usually 2–5 m above ground level.

Silviculture   The cultivation of trees for commercial purposes.

Singling   The selective cutting of a coppice leaving one stem which is encouraged to grow as a future timber tree.

Softwood   The wood of coniferous trees (which may or may not be soft).

SSSI   Site of special scientific interest, a designation given by English Nature to sites of nature conservation importance.

Stand   A group of trees, often of similar age and species.

Standard   A tree which has a single stem. Often applied to the trees not coppiced in a coppice block.

Stool   The part of a coppice which is left after cutting. Sometimes also the result of natural events such as wind damage.

Stored coppice   A single coppice branch which has been retained on the stool for the production of timber trees.

Succession   The sequence of colonisation by trees, and other species, of land which is open.

Sucker   New growth of a tree arising from the roots or an underground stem.

Sun scorch   Burning of leaves or trunk by the sun, usually because the tree, or plant, has recently been exposed to more sunlight than it is used to.

Symbiotic   A relationship between two organisms where both benefit.

Thinning   The removal of selected trees (or branches) to give those remaining more light and room to grow.

Timber   The woody part of a tree which can be cut into large pieces of wood.

Tree   A woody plant able to exceed 6 m in height and having a woody, usually single stem.

Understorey   The crowns of any trees growing underneath the main canopy of the wood.

Underwood   The woody component of the shrub layer (often applied to coppice). Also used for the cut wood.

Wildwood   The type of woodland which is presumed to have occurred in the UK after the retreat of the ice sheet and before the intervention of humans.

Woodland   Land covered mainly by trees.

Wood-pasture   Woodland which is grazed.

SPECIES LIST

•

This section lists the species mentioned in the book, drawing them together into their appropriate taxonomic groupings. Organisms (except viruses) are first divided into one of five kingdoms (Table 6.1). There are several more subdivisions: the major levels used here are shown in Table 6.2. The final division is into species. The naming of organisms in a standard way (following international convention) enables everyone to be certain which species is under consideration.

Every species is identified by a unique scientific name consisting of a binomial term (two words, the first is the genus and the second is the species). Sometimes, especially where there are several species that are difficult to separate, a different term (e.g. agg. or sect.) is placed after the genus, to indicate that several species are involved. In other cases, the name may include an x indicating that the organism (usually a plant) is a cross between two different species. A third term may be used to designate a sub-species or a variety of the species. Note that species and sub-species names

Table 6.1: Summary of the five kingdoms

Kingdom	Organisms	Characteristics
Prokaryotae (Monera)	Bacteria and Cyanobacteria	Single-celled, prokaryotic (lack a membrane-bounded nucleus)
Protoctista	Nucleated algae (including seaweeds), protozoa and slime moulds	Those eukaryotes (i.e. possess a membrane-bounded nucleus) which are not fungi, plants or animals. Often single-celled, mainly aquatic (including in damp environments and the tissues of other species), often autotrophic
Fungi	Fungi and lichens	Eukaryotic, mainly multicellular, develop directly from spores with no embryological development, heterotrophic and often saprotrophic (feed on non-living organic matter)
Plantae	Plants including mosses, ferns, liverworts, conifers and flowering plants	Eukaryotic, multicellular, develop from an embryo (multicellular young organisms supported by maternal tissue), often photoautotrophic (autotrophs obtaining energy from sunlight)
Animalia	Invertebrate and vertebrate animals	Eukaryotic, multicellular, develop from a blastula (hollow ball of cells), heterotrophic

Source: Margulis and Schwartz (1988)

Table 6.2: Taxonomic levels used in species list

....

KINGDOM
PHYLUM
CLASS
   Order
      Family
         Species Authority [Common Name]

....

are Latinised and should be shown in italics, with the generic name beginning with a capital letter. The species name is followed by the name of the person who originally described it (the authority). Where the authority is well known (e.g. if they have named many species), an abbreviation may be used (e.g. L. for Linnaeus, F. for Fabricius). An authority in parentheses indicates that the original species name has been altered (e.g. if the organism has been placed in a different genus). In the absence of parentheses around the authority, the original name is still in use today.

The names used here follow a variety of sources depending on the taxonomic group: Hill, Preston and Smith (1991–94) for mosses and liverworts; Stace (1997) for plants; Roberts (1985) for spiders; Kloet and Hincks (1964–75) for insects; Cramp (1977–94) for birds; Corbet and Harris (1991) for mammals.

PROKARYOTAE
NITROGEN-FIXING AEROBIC BACTERIA
PROTOCTISTA
GYMNOMYCOTA
MYXOMYCETES
   Slime moulds
OOMYCOTA
OOMYCETES
   Pythiales
      Pythiacea
         Phytophthora de Bary species
FUNGI
BASIDIOMYCOTA (Mushrooms and Toadstools)
HYMENOMYCETES
   Agaricales
      Amanitaceae
         Amanita muscaria (L.) [Fly Agaric]
      Tricholomataceae
         Laccaria amethystea (Bull.) [Amethyst Deceiver]
         Armillaria gallica Marxmüller and Romagn.
         Collybia (Fr.) species
   Aphyllophorales
      Corticiaceae
         Peniophora Cooke species
      Polyporaceae
         Heterobasidion annosum (Fr.) [Root Fomes]

         Piptoporus betulinus (Bull.) [Birch Polypore]
         Laetiporus sulphureus (Bull. ex Fr) [Chicken of the Woods]
ASCOMYCOTA
PYRENOMYCETES
   Ophiostomatales
      Ophiostomataceae
         Ceratocystis ulmi [Elm Bark Disease Fungus]
   Sphaeriales
      Xylariaceae
         Xylaria hypoxylon (L.) [Candle Snuff Fungus]
MYCOPHYCOPHYTA (Lichens)
ASCOMYCETES
   Lecanorales
      Lobariaceae
         Lobaria pulmonaria (L.) [Lung wort]
PLANTAE (plants)
BRYOPHYTA (Mosses and Liverworts)
MUSCI (Mosses)
   Sphagnales
      Sphagnaceae
         Sphagnum L. species
         Sphagnum capillifolium (Ehrh.)
   Dicranales
      Dicranaceae
         Dicranum majus Sm.

Leucobryaceae
Leucobryum glaucum (Hedw.) [Cushion moss]
Thuidiales
Thuidium tamariscinum (Hedw.)
Hypnobryales
Brachytheciaceae
Isothecium myosuroides Brid.
Hypnaceae
Hylocomium splendens (Hedw.)
Rhytidiadelphus triquetrus (Hedw.)
HEPATICAE (Liverworts)
Jungermannineae
Scapaniaceae
Diplophyllum albicans (L.)
Scapania gracilis Lindb.
SPHENOPHYTA
EQUISETOPSIDA (Horsetails)
Equisetaceae (Horsetail family)
FILICINOPHYTA (Ferns)
PTEROPSIDA
Dennstaedtiaceae (Bracken family)
Pteridium aquilinum (L.) [Bracken]
Dryopteridaceae (Buckler-fern family)
Dryopteris dilatata (Hoffm.) [Broad Buckler-fern]
Dryopteris filix-mas (L.) [Male-fern]
Blechnaceae (Hard-fern family)
Blechnum spicant (L.) [Hard Fern]
CONIFEROPHYTA (Conifers)
PINOPSIDA
Pinaceae (Pine family)
Abies alba Miller [European Silver Fir]
Abies grandis (Douglas) [Giant Fir]
Abies procera Rehder [Noble Fir]
Pseudotsuga menziesii (Mirbel) [Douglas Fir]
Tsuga heterophylla (Raf.) [Western Hemlock-Spruce]
Picea sitchensis (Bong.) [Sitka Spruce]
Picea omorika (Pancic) [Serbian Spruce]
Picea abies (L.) [Norway Spruce]
Larix decidua Miller [European Larch]
Larix kaempferi (Lindley) [Japanese Larch]
Cedrus libani A. Rich. [Cedar-of-Lebanon]
Pinus sylvestris L. [Scots Pine]
Pinus sylvestris scotica (P.K. Schott) [Caledonian Pine]
Pinus pinea L. [Stone Pine]
Pinus nigra laricio Maire [Corsican Pine]

Pinus contorta Douglas [Lodgepole Pine]
Pinus pinaster Aiton [Maritime Pine]
Taxodiaceae (Redwood family)
Sequoiadendron giganteum (Lindley) [Wellingtonia]
Taxodium distichum (L.) [Swamp Cypress]
Metasequoia glyptostroboides Hu and W.C. Cheng [Dawn Redwood]
Cupressaceae (Juniper family)
Thuja plicata Donn [Western Red-cedar]
Juniperus communis L. [Common Juniper]
Araucariaceae (Monkey-puzzle family)
Araucaria araucana (Molina) [Monkey-puzzle]
Taxaceae (Yew family)
Taxus baccata L. [Yew]
ANGIOSPERMOPHYTA (Flowering Plants)
MAGNOLIOPSIDA (Dicotyledons)
Ranunculales
Ranunculaceae (Buttercup family)
Anenome nemorosa L. [Wood Anenome]
Ranunculus ficaria L. [Lesser Celandine]
Ranunculus repens L. [Creeping Buttercup]
Ranunculus auricomus L. [Goldilocks Buttercup]
Hamamelidales
Platanaceae (Plane family)
Platanus x hispanica Mill. ex Muenchh. [London plane]
Urticales
Ulmaceae (Elm family)
Ulmus glabra Huds. [Wych Elm]
Ulmus x hollandica Miller [Dutch Elm]
Ulmus procera Salisb. [English Elm]
Ulmus minor angustifolia (Weston) [Cornish Elm]
Cannabaceae (Hop family)
Humulus lupulus L. [Hop]
Urticaceae (Nettle family)
Urtica dioica L. [Common Nettle]
Juglandales
Juglandaceae (Walnut family)
Juglans regia L. [Walnut]
Juglans nigra L. [Black Walnut]
Pterocarya Kunth species [Wingnuts]
Fagales
Fagaceae (Beech family)
Fagus sylvatica L. [Beech]
Nothofagus Blume species [Southern Beeches]

Nothofagus obliqua (Mirbel) [Roble]
Nothofagus nervosa (Philippi) [Rauli]
Castanea sativa Mill. [Sweet Chestnut]
Quercus cerris L. [Turkey Oak]
Quercus ilex L. [Evergreen Oak]
Quercus petraea (Matt.) [Sessile Oak]
Quercus robur L. [Pedunculate Oak]
Quercus rubra L. [Red Oak]
Quercus coccinea Muenchh. [Scarlet Oak]
Quercus suber L. [Cork Oak]

Betulaceae (Birch family)
Betula pendula Roth [Silver Birch]
Betula pubescens Ehrh. [Downy Birch]
Betula nana L. [Dwarf Birch]
Alnus glutinosa (L.) [Alder]
Alnus incana (L.) [Grey Alder]
Alnus cordata (Lois.) [Italian Alder]
Carpinus betulus L. [Hornbeam]
Corylus avellana L. [Hazel]

Caryophyllales
Caryophyllaceae (Pink family)
Stellaria neglecta Weihe [Greater Chickweed]

Theales
Clusiaceae (St John's-wort family)
Hypericum pulchrum L. [Slender St John's-wort]

Malvales
Tiliaceae (Lime family)
Tilia platyphyllos Scop. [Large-leaved Lime]
Tilia cordata Mill. [Small-leaved Lime]
Tilia x europaea L. [Lime]

Droseraceae (Sundew family)
Drosera intermedia Hayne [Oblong-leaved Sundew]

Violales
Violaceae (Violet family)
Viola riviniana Rchb. [Common Dog-violet]
Viola reichenbachiana Jordan [Early Dog-violet]

Salicales
Salicaceae (Willow family)
Populus alba L. [White Poplar]
Populus x canescens (Aiton) [Grey Poplar]
Populus tremula L. [Aspen]
Populus nigra L. [Black-poplar]
Populus nigra var. *italica* Moench [Lombardy Poplar]
Populus x canadensis Moench [Hybrid Black Poplar]

Populus trichocarpa Torrey and A. Gray [Western Balsam-poplar]
Salix pentandra L. [Bay Willow]
Salix fragilis L. [Crack Willow]
Salix alba L. [White Willow]
Salix alba caerulea (Smith) [Cricket-bat Willow]
Salix triandra L. [Almond Willow]
Salix viminalis L. [Osier]
Salix caprea L. [Goat Willow]
Salix cinerea L. [Grey Willow]

Capperales
Brassicaceae (Cabbage and cress family)

Ericales
Ericaceae (Heather family)
Rhododendron ponticum L. [Rhododendron]
Andromeda polifolia L. [Bog-rosemary]
Calluna vulgaris (L.) [Heather]
Erica tetralix L. [Cross-leaves heath]
Erica cinerea L. [Bell Heather]
Vaccinium vitis-idaea L. [Cowberry]
Vaccinium myrtillus L. [Bilberry]

Monotropaceae (Bird's-nest family)
Monotropa hypopitys L. [Yellow Bird's-nest]

Primulales
Primulaceae (Primrose family)
Primula vulgaris Huds. [Primrose]
Primula elatior (L.) [Oxlip]
Lysimachia vulgaris L. [Yellow Loosestrife]
Anagallis arvensis arvensis L. [Scarlet Pimpernel]

Grossulariaceae (Gooseberry family)
Ribes L. species [Gooseberries and currants]

Rosales
Saxifragaceae (Saxifrage family)
Chrysosplenum oppositifolium L. [Opposite-leaved Golden Saxifrage]

Rosaceae (Rose family)
Filipendula ulmaria (L.) [Meadowsweet]
Rubus sect. *Glandulosus* Wimm. and Grab. [Brambles]
Geum urbanum L. [Wood Avens]
Prunus spinosa L. [Blackthorn]
Prunus avium (L.) [Wild Cherry]
Prunus padus L. [Bird Cherry]
Pyrus pyraster (L.) [Wild Pear]
Malus sylvestris (L.) [Crab Apple]
Malus domestica Borkh. [Apple]
Sorbus aucuparia L. [Rowan]

Sorbus aria (L.) [Common Whitebeam]
Sorbus intermedia (Ehrh.) [Swedish Whitebeam]
Sorbus torminalis (L.) [Wild Service Tree]
Crataegus monogyna Jacq. [Hawthorn]
Crataegus laevigata (Poir.) [Midland Hawthorn]
Fabaceae (Pea family)
Robinia pseudoacacia L. [False-acacia]
Laburnum anagyroides Medikus [Laburnum]
Myrtales
Myrtaceae (Myrtle family)
Eucalyptus L'Her. [Gums]
Onagraceae (Willowherb family)
Chamerion angustifolium (L.) [Rosebay Willowherb]
Circaea lutetiana L. [Enchanter's-nightshade]
Celastrales
Aquifoliaceae (Holly family)
Ilex aquifolium L. [Holly]
Euphorbiales
Euphorbiaceae (Spurge family)
Mercurialis perennis L. [Dog's Mercury]
Euphorbia amygdaloides L. [Wood Spurge]
Rhamnaceae (Buckthorn family)
Frangula alnus Miller [Alder Buckthorn]
Sapindales
Hippocastanaceae (Horse-chestnut family)
Aesculus hippocastanum L. [Horse-chestnut]
Aceraceae (Maple family)
Acer platanoides L. [Norway Maple]
Acer campestre L. [Field Maple]
Acer pseudoplatanus L. [Sycamore]
Geraniales
Oxalidaceae (Wood-sorrel family)
Oxalis acetosella L. [Wood-sorrel]
Geraniaceae (Crane's-bill family)
Geranium robertianum L. [Herb-robert]
Apiales
Araliaceae (Ivy family)
Hedera helix L. [Ivy]
Apiaceae (Carrot family)
Sanicula europaea L. [Sanicle]
Conopodium majus (Gouan) [Pignut]
Lamiales
Lamiaceae (Deadnettle family)
Lamiastrum galeobdolon (L.) [Yellow Archangel]
Teucrium scorodonia L. [Wood Sage]
Glechoma hederacea L. [Ground-ivy]
Plantaginaceae (Plantains)
Plantago L. species [Plantains]
Scrophulariales
Oleraceae (Ash family)
Fraxinus excelsior L. [Ash]
Scrophulariaceae (Figwort family)
Digitalis purpurea L. [Foxglove]
Veronica chamaedrys L. [Germander Speedwell]
Veronica montana L. [Wood Speedwell]
Melampyrum L. Species [Cow-wheats]
Lathraea (Toothworts)
Lathraea squamaria L. [Toothwort]
Campanulaceae (Bellflower family)
Campanula latifolia L. [Giant Bellflower]
Rubiales
Rubiaceae (Bedstraw family)
Galium aparine L. [Cleavers]
Dipsacales
Caprifoliaceae (Honeysuckle family)
Sambucus nigra L. [Elder]
Viburnum opulus L. [Guelder-rose]
Linnaea borealis L. [Twinflower]
Lonicera periclymenum L. [Honeysuckle]
Adoxaceae (Moschatel family)
Adoxa moschatellina L. [Moschatel]
Valerianaceae (Valerian family)
Valariana officinalis L. [Common Valerian]
Valariana dioica L. [Marsh Valarian]
Asterales
Asteraceae (Daisy family)
Cirsium palustre (L.) [Marsh Thistle]
Crepis paludosa (L.) [Marsh Hawk's-beard]
LILIIDAE (Monocotyledons)
Araceae (Lords-and-ladies family)
Arum maculatum L. [Lords-and-ladies]
Juncales
Juncaceae (Rush family)
Juncus effusus L. [Soft Rush]
Luzula pilosa (L.) [Hairy Wood-rush]
Cyperales
Cyperaceae (Sedge family)
Carex paniculata L. [Greater Tussock-sedge]
Carex remota L. [Remote Sedge]
Carex rostrata Stokes [Bottle Sedge]
Carex pendula Hudson [Pendulous Sedge]

Carex sylvatica Hudson [Wood Sedge]

Poaceae (Grass family)

Festuca rubra L. agg. [Red Fescue]

Deschampsia cespitosa (L.) [Tufted Hair-grass]

Deschampsia flexuosa (L.) [Wavy Hair-grass]

Holcus lanatus L. [Yorkshire-fog]

Holcus mollis L. [Creeping Soft-grass]

Anthoxanthum odoratum L. [Sweet Vernal Grass]

Agrostis capillaris L. [Common Bent]

Agrostis canina L. [Velvet Bent]

Calamagrostis canescens (Wigg.) [Purple Small-reed]

Brachypodium sylvaticum (Huds.) [False Brome]

Molinia caerulea (L.) [Purple Moor-grass]

Phragmites australis (Cav.) [Common Reed]

Liliales

Liliaceae (Lily family)

Narthecium ossifragum (L.) [Bog Asphodel]

Convallaria majalis L. [Lily-of-the-valley]

Paris quadrifolia L. [Herb Paris]

Hyacinthoides non-scripta (L.) [Bluebell]

Allium ursinum L. [Ransoms]

Narcissus pseudonarcissus L. [Daffodil]

Orchidales

Orchidaceae (Orchid family)

Goodyera repens (L.) [Creeping Lady's-tresses]

Orchis mascula (L.) [Early-purple Orchid]

ANIMALIA (animals)

NEMATODA (Nematode worms or Roundworms)

ANNELIDA (Segmented Worms)

OLIGOCHAETA

Haplotaxida

Lumbricidae (Earthworms)

ARTHROPODA (Arthropods)

MALACOSTRACA

Isopoda (Woodlice)

ARACHNIDA

Acarina (Mites and Ticks)

Pseudoscorpiones (Pseudoscorpions)

Opiliones (Harvestmen)

Araneae (Spiders)

Theriididae

Theridion pallens Blackwall

Linyphiidae

Drapetisca socialis (Sundevall)

DIPLOPODA (Millipedes)

Julidae

Cylindroiulus punctatus (Leach)

CHILOPODA (Centipedes)

INSECTA (Insects)

Collembola (Springtails)

Orthoptera (Grasshoppers and crickets)

Tettigoniidae

Meconema thalassinum (De Geer) [Oak Bush Cricket]

Coleoptera (Beetles)

Buprestidae (Jewel Beetles)

Agrilus pannonicus [Oak Jewel Beetle]

Lucanidae (Stag Beetles)

Lucanus cervus L. [Stag Beetle]

Scolytidae (Bark and Ambrosia Beetles)

Scolytus Müller species

Lepidoptera (Butterflies and Moths)

Pieridae

Gonepteryx rhamni (L.) [Brimstone]

Riodinidae

Hamearis lucina (L.) [Duke of Burgundy]

Nymphalidae

Ladoga camilla (L.) [White Admiral]

Apatura iris (L.) [Purple Emperor]

Polygonia c-album (L.) [Comma]

Boloria eurhrosyne (L.) [Pearl-bordered Fritillary]

Argynnis adippe (L.) [High Brown Fritillary]

Argynnis paphia (L.) [Silver-washed Fritillary]

Mellicta athalia Rottemburg [Heath Fritillary]

Satyridae (Brown Butterflies)

Pararge aegeria (L.) [Speckled Wood]

Pyronia tithonius (L.) [Gatekeeper]

Maniola jurtina (L.) [Meadow Brown]

Aphantopus hyperantus (L.) [Ringlet]

Lycaenidae (Blue Butterflies and Allies)

Thecla betulae L. [Brown Hairstreak]

Quercusia quercus (L.) [Purple Hairstreak]

Strymondidia w-album Knock [White-letter Hairstreak]

Celastrina argiolus L. [Holly Blue]

Hesperiidae (Skippers)

Thymelicus flavus Poda [Small Skipper]

Hepialidae

Hepialus lupulinus L. [Common Swift]

Thyatiridae

Cymatophorima diluta hartwiegi Reisser [Oak Lutestring]

Polyploca ridens F. [Frosted Green]
Geometridae
Cyclophora punctaria Linnaeus [Maiden's Blush]
Idaea biselata Hufnagel [Small Fan-footed wave]
Idaea aversata L. [Riband Wave]
Hydriomena furcata Thunberg [July Highflyer]
Rheumaptera undulata L. [Scallop Shell]
Operophtera brumata L. [Winter moth]
Epirrita dilutata Denis and Schiffermuller [November Moth]
Colotois pennaria L. [Feathered Thorn]
Lycia hirtaria Cl. [Brindled Beauty]
Agriopis leucophaearia Denis and Schiffermuller [Spring Usher]
Agriopis aurantiaria Hb. [Scarce Umber]
Agriopis marginaria F. [Dotted Border]
Erannis defoliaria Cl. [Mottled Umber]
Peribatodes rhomboidaria Denis and Schiffermuller [Willow Beauty]
Serraca punctinalis Scop. [Pale Oak Beauty]
Ectropis bistortata Goeze. [The Engrailed]
Alcis repandata L. [Mottled Beauty]
Sphingidae (Hawkmoths)
Deilephila elpenor (L.) [Elephant Hawk-moth]
Laothoe populi Linaeus [Poplar Hawk]
Notodontidae
Peridea anceps Goeze [Great Prominent]
Clostera curtula L. [Chocolate Tip]
Diloba caeruleocephala L. [Figure of Eight]
Noctuidae
Agrotis exclamationis L. [Heart and Dart]
Noctua pronuba L. [Large Yellow Underwing]
Diarsia mendica F. [Ingrailed Clay]
Panolis flammea Denis and Schiffermuller [Pine Beauty]
Orthosia cruda Denis and Schiffermuller [Small Quaker]
Orthosia stabilis Denis and Schiffermuller [Common Quaker]
Orthosia incerta Hufnagel [Clouded Drab]
Orthosia gothica L. [Hebrew Character]
Brachionycha sphinx Hufnagel [The Sprawler]
Allophyes oxyacanthae L. [Green-brindled Crescent]

Agrochola circellaris Hufn. [The Brick]
Conistra rubiginea Denis and Schiffermuller [The Chestnut]
Amphipyra pyramidea L. [Copper Underwing]
Rusina ferruginea Esp. [Brown Rustic]
Cosmia trapezina L. [The Dun-bar]
Apamea monoglypha Hufnagel [Dark arches]
Oligia strigilis Linnaeus [Marbled Minor]
Photedes minima Haw. [Small Dotted Buff]
Diachrysia chrysitis L. [Burnished Brass]
Autographa gamma L. [Silver Y]
Hypena proboscidalis L. [The Snout]
Diptera
Mycetophilidae (Fungus gnats)
Hymenoptera
Cynipidae
Neuroterus numismalis (Geoffroy)
Neuroterus quercusbaccarum (L.)
Neuroterus albipes (Schenck)
Neuroterus tricolor (Hartig)

CHORDATA
AVES (Birds)
Accipitriformes
Accipitridae (Hawks and Allies)
Accipiter nisus (L.) [Sparrowhawk]
Galliformes
Phasianidae (Partridges, Pheasants and Allies)
Phasianus colchicus L. [Pheasant]
Charadiiformes (Plovers and Allies)
Charadriidae (Plovers and Lapwings)
Numenius arquata L. [Curlew]
Scolopacidae (Sandpipers and Allies)
Gallinago gallinago (L.) [Snipe]
Scolopax rusticola L. [Woodcock]
Columiformes
Columbidae (Pigeons and Doves)
Columba palumbus L. [Wood pigeon]
Streptopelia turtur (L.) [Turtle Dove]
Stringiformes
Strigidae (Brown Owls and Allies)
Strix aluco L. [Tawny Owl]
Caprimulgiformes
Caprimulgidae (Nightjars)
Caprimulgus europaeus Von Linne [Nightjar]
Piciformes
Picidae (Woodpeckers and Allies)
Picus viridis (L.) [Green Woodpecker]
Dendrocops major (L.) [Great-spotted

Woodpecker]
Dendrocops minor (L.) [Lesser-spotted Woodpecker]
Passeriformes (Perching Birds)
Alaudidae (Larks)
Lullula arborea (L.) [Woodlark]
Motacillidae (Pipits and Wagtails)
Anthus trivialis (L.) [Tree Pipit]
Anthus pratensis (L.) [Meadow Pipit]
Troglodytidae (Wrens)
Troglodytes troglodytes (L.) [Wren]
Sylviidae (Warblers and Allies)
Sylvia borin (Boddaert) [Garden Warbler]
Sylvia atricapilla (L.) [Blackcap]
Phylloscopus trochilus (L.) [Willow Warbler]
Phylloscopus collybita (Viellot) [Chiffchaff]
Phylloscopus sibilatrix (Bechstein) [Wood Warbler]
Regulus regulus (L.) [Goldcrest]
Muscicapidae (Flycatchers, Warblers, etc.)
Ficedula hypoleuca Pallas [Pied Flycatcher]
Muscicapa striata (Pallas) [Spotted Flycatcher]
Phoenicurus phoenicurus (L.) [Redstart]
Erithacus rubecula (L.) [Robin]
Luscinia megarhynchos Brehm [Nightingale]
Turdus merula L. [Blackbird]
Aegithalidae (Long-tailed Tits and Allies)
Aegithalos caudatus (L.) [Long-tailed Tit]
Paridae (Tits)
Parus palustris L. [Marsh Tit]
Parus cristatus L. [Crested Tit]
Parus ater L. [Coal Tit]
Parus caeruleus L. [Blue tit]
Parus major L. [Great tit]
Sittidae (Nuthatches)
Sitta europaea L. [Nuthatch]
Certhiidae (Treecreepers)
Certhia familiaris L. [Treecreeper]
Corvidae (Crows and Allies)
Garrulus glandarius (L.) [Jay]
Fringillidae (Finches)
Fringilla coelebs L. [Chaffinch]
Coccothraustes coccothraustes L. [Hawfinch]
Carduelis spinus (L.) [Siskin]
Frinillidae (Crossbills)

Loxia curvirostra L. [Crossbill]
Loxia scotica (Hartert) [Scottish Crossbill]
MAMMALIA (Mammals)
Insectivora (Insectivores)
Soricidae (Shrews)
Sorex araneus L. [Common Shrew]
Sorex minutus L. [Pygmy Shrew]
Neomys fodiens (Pennant) [Water Shrew]
Chiroptera (Bats)
Vespertilionidae
Myotis bechsteinii (Kuhl) [Bechstein's Bat]
Myotis daubentoni (Kuhl) [Daubenton's Bat]
Nyctalus noctula (Schreber) [Noctule Bat]
Pipistrellus pipistrellus (Shreber) [Pipistrelle Bat]
Rodentia (Rodents)
Castor fiber L. [Beaver]
Sciuridae (Squirrels)
Sciurus vulgaris L. [Red Squirrel]
Sciurus carolinensis Gmelin [Grey Squirrel]
Muridae (Voles, Rats and Mice)
Clethrionomys glareolus Schreber [Bank Vole]
Microtus agrestis L. [Field Vole]
Apodemus sylvaticus (L.) [Wood Mouse]
Apodemus flavicollis Melchior [Yellow-necked Mouse]
Micromys minutus (Pallas) [Harvest Mouse]
Gliridae (Dormice)
Muscardinus avellanarius L. [Common Dormouse]
Glis glis (L.) [Edible Dormouse]
Carnivora (Terrestrial Carnivores)
Mustelidae (Weasels and Allies)
Mustela erminea L. [Stoat]
Mustela nivalis L. [Weasel]
Artiodactyla
Suidae
Sus scrofa L. [Wild Boar]
Cervidae (Deer)
Cervus elaphus L. [Red Deer]
Cervus nippon Temminck [Sika Deer]
Dama dama (L.) [Fallow Deer]
Capreolus capreolus (L.) [Roe Deer]
Muntiacus reevesi (Ogilby) [Muntjac]
Hydropotes inermis Swinhoe [Chinese Water Deer]

FURTHER READING

•

The list below gives some of the major texts that assist in the study of the habitats discussed in this book. They are listed under sections dealing with practical techniques (including the identification of plants and animals), practical conservation, new initiatives and journals.

PRACTICAL TECHNIQUES

Several texts help with the design of experiments and methods of approaching a particular ecological problem.

Chalmers, N. and Parker, P. (1989) *The Open University Project Guide*, 2nd edition, Field Studies Occasional Publications No. 9, Milton Keynes: Open University Press.

Gilbertson, D. D., Kent, M. and Pyatt, F. B. (1985) *Practical Ecology for Geography and Biology*, London: Unwin Hyman.

Williams, G. (1987) *Techniques and Field Work in Ecology*, London: Bell & Hyman Ltd.

Data analysis

It is important to incorporate statistical techniques into the experimental design, since a poorly designed experiment can be difficult to interpret. There are a number of reference texts, but most are quite heavy-going. The following are a few of the more user-friendly student texts.

Chalmers, N. and Parker, P. (1989) *The Open University Project Guide*, 2nd edition., Field Studies Occasional Publications No. 9, Milton Keynes: Open University Press.

Ebden, D. (1987) *Statistics in Geography*, Oxford: Blackwell.

Fowler, J. and Cohen, L. (1990) *Practical Statistics for Field Biology*, Milton Keynes: Open University Press.

Kent, M. and Coker, P. (1992) *Vegetation Description and Analysis*, Chichester: John Wiley & Sons Ltd.

Watt, T. A. (1993) *Introductory Statistics for Biology Students*, London: Chapman & Hall.

Identification

There are two types of identification guide. Some are descriptive and usually contain colour illustrations. Care should be taken when using descriptive guides since it is easy to confuse superficially similar species. Better are those texts that incorporate keys (where organisms are sequentially separated out using diagnostic characters).

Field guides are accessible descriptive guides to either specific groups of plants or animals, or to particular habitats. HarperCollins and Countrylife publish mainly descriptive guides, while Warne keys to wildflowers, birds and trees provide an alternative which contain identification keys. Field Studies Council AIDGAP keys are user-friendly and cover a wide range of plant and animal groups, especially invertebrates. Naturalists' Handbooks

(Richmond Publishing Company) are other user friendly keys to either specific groups of invertebrate animals (mainly insects), or to the occupants of particular habitats. Other identification texts exist which are not part of a series (e.g. Skinner, 1984; Marshall and Haes, 1988).

More specialist keys are aimed towards the professional and, although they may be difficult for the beginner to use, they are usually more complete than the examples given above. Specialist keys include those produced by the Freshwater Biological Association for the identification of British freshwater invertebrates, the Linnean Society Synopses of the British Fauna covering a large number of invertebrate groups (e.g. earthworms, harvestmen, woodlice, millipedes), and the Royal Entomological Society of London handbooks for the identification of British insects. Other texts such as Stace (1997) for identifying plants do not form part of a series.

PRACTICAL CONSERVATION

Tait, J., Lane, A. and Carr, S. (1988) *Practical Conservation: Site Assessment and Management Planning*, Kent: Hodder and Stoughton.
BTCV (1997) *Woodlands: A Practical Handbook*, Reading: BTCV.

NEW INITIATIVES

Community Forests have information available regarding their current projects, as do local naturalist's trusts and the organisations responsible for the management of local nature reserves. The Forestry Authority and Forest Enterprise produce regular reports of their work. English Nature publishes leaflets relating to the management of woodlands and the Countryside Commission has a wide range of publications concerning countryside issues.

JOURNALS

Several journals cover ecology, management and conservation issues including those related to woodland areas, for example:

Arboricultural Journal
Biological Conservation
British Wildlife
Environmental Management
Forestry and British Timber
Journal of Applied Ecology
Journal of Environmental Management
The Journal of Practical Ecology and Conservation
Quarterly Journal of Forestry.

English Nature produces a magazine (*Enact*) which covers a variety of habitat management issues, techniques and case studies. The Tree Council publishes a magazine (*Tree News*) which includes topical articles relating to trees and woodland.

Local natural history journals give details of surveys and sites of local importance, for example:

Essex Naturalist
Lancashire Wildlife Journal
The London Naturalist
The Naturalist (North of England)
North West Naturalist
Proceedings of the Bristol Naturalists' Society
Sorby Record (covering Sheffield).

REFERENCES

•

Aldous, J.R. (1997) British Forestry: 70 years of achievement. *Forestry*, 70 (4): 283–291.

Alexander, K., Green, E.E., and Key, R. (1996) The management of overmature tree populations for nature conservation: the basic guidelines. In H.J. Read (ed.) *Pollard and Veteran Tree Management II*, London: Corporation of London, pp. 122–135.

Alexander, K., Green, T., and Key, R. (1998) Managing our ancient trees. *Tree News*, Spring: 10–13.

Archer, P., Fordham, P. and Harding, M. (1995) *Bradfield Woods: Management Plan*, unpublished manuscript.

Askew, R.R. and Redfern, M. (1992) *Plant Galls*, Slough: Richmond Publishing Co. Ltd.

Austad, I. (1988) Tree pollarding in Western Norway. In H.H. Birks, H.J.B. Birks, P.E. Kaland and D. Moe (eds) *The Cultural Landscape: Past, Present and Future*, Cambridge: Cambridge University Press.

Bang, P. and Dahlstrom, P. (1980) *Animal Tracks and Signs*, London: Collins.

Banks, W.B. and Cooper, R.J. (1997) Utilization of softwoods in Great Britain. *Forestry*, 70 (4): 315–318.

Barwick, P. (1996) The Birklands Oak Project. In H.J. Read (ed.) *Pollard and Veteran Tree Management II*, pp. 69–70.

Battell, G. (1996) Our ancient trees: the way ahead. In H.J. Read (ed.) *Pollard and Veteran Tree Management II*, London: Corporation of London.

Bealey, C.E. and Robertson, P.A. (1992) Coppice management for pheasants. In G.P. Buckley (ed.) *Ecology and Management of Coppice Woodland*, London: Chapman and Hall.

Begon, M., Harper, J.L. and Townsend, C.R. (1996) *Ecology*, 3rd edition, Oxford: Blackwell Scientific Publications.

Bell, S. (1993) How fares the National Forest? *Quarterly Journal of Forestry* 87: 124–128.

Bellamy, D. (1972) *Bellamy on Botany*, London: BBC.

Bellmann, H. (1985) *A Field Guide to the Grasshoppers and Crickets of Britain and Northern Europe*, London: Collins.

Bibby, C.J., Phillips, B.N. and Seddon, A.J. (1985) Birds of restocked conifer plantations in Wales. *Journal of Applied Ecology*, 22: 619–633.

Bignal, E., Bratton, J., Fuller, R., Pienkowski, M., Tubbs, C. and Warren, M. (1995) Letter in reply to Hambler and Speight 1995a. *British Wildlife*, 6 (5): 337–338.

Blower, J.G.B. (1985) *Millipedes*. Synopses of the British Fauna (New Series) No. 35. London: E.J. Brill/Dr W. Backhuys.

Bowden, C. and Hoblyn, R. (1990) The increasing importance of restocked conifer plantations for woodlarks in Britain: implications and consequences. *RSPB Conservation Review*, 4: 1–17.

Brown, A. (1992) *The U.K. Environment*, London: HMSO.

Brown, N. (1997) Re-defining native woodland. *Forestry*, 70 (3): 191–198.

Brown, R.W., Lawrence, M.J. and Pope, J. (1984) *Animals of Britain and Europe: Their Tracks, Trails and Signs*, Middlesex: Countrylife Books.

BTCV (1997) *Woodlands: A Practical Handbook*, Reading: BTCV.

Bunce, R.G.H. (1982) *A Field Key for Classifying British Woodland Vegetation. Part 1*, Cambridge: Institute of Terrestrial Ecology.

Carroll, J. and Robertson, P. (1997) *Integrating Pheasant Management and Woodland Conservation*, Fordingbridge: Game Conservancy Council.

Chinery, M. (1976) *A Field Guide to the Insects of Britain and Northern Europe*, 2nd edition, London: Collins.

Clouston, B. and Stansfield, K. (1979) *After the Elm*, London: Heinemann.

Colquhoun, M.K. and Morley, A. (1943) Vertical

zonation in woodland bird communities. *Journal of Animal Ecology*, 12: 75–81.

Corbet, G.B. and Harris, S. (eds) (1991) *The Handbook of British Mammals*, 3rd edition, Oxford: Blackwell Scientific Publications.

Countryside Commission (1991) *After the Great Storm*, Cheltenham: Countryside Commission.

Countryside Commission (1992) *The Chilterns Landscape*, Cheltenham: Countryside Commission.

Countryside Commission (1993a) *Rooting for the Future*, Cheltenham: Countryside Commission.

Countryside Commission (1993b) *England's Trees and Woods*, Cheltenham: Countryside Commission.

Countryside Commission (1995) *Growing in Confidence: Understanding People's Perceptions of Urban Fringe Woodlands*, Cheltenham: Countryside Commission.

Cramp, S. (1977–1994) *Handbook of the Birds of Europe, the Middle East and North Africa: The Birds of the Western Palearctic*, Volumes 1, 2, 3, 4, 5 and 8. Oxford: Oxford University Press.

Davis, B., Walker, N., Ball, D and Fitter, A. (1992) *The Soil*, London: HarperCollins.

Department of the Environment (1993) *Air Pollution and Tree Health in the United Kingdom*, London: HMSO.

Didham, R.K., Ghazoul, J., Stork, N.E. and Davis, A.J. (1996) Insects in fragmented forests: a functional approach. *TREE*, 11 (6): 255–260.

Dixon, R.K, Brown, S., Houghton, R.A., Solomon, A.M., Trexeler, M.C. and Wisniewski, J. (1994) Carbon pools and flux of global forest ecosystems. *Science*, 263: 185–190.

Dobson, F. (1992) *Lichens*, 3rd edition, Slough: Richmond Publishing Co. Ltd.

Edwards, R. (1997) Return of the wee trees. *New Scientist*, 11 October 1997: 15.

English Nature (1994) *Species Conservation Handbook*, Peterborough: English Nature.

English Nature (1996) *Guidelines for Identifying Ancient Woodlands*, Peterborough: English Nature.

English Nature (1997) *Deer Management and Woodland Conservation in England*, Peterborough: English Nature.

English Nature (1998) *Management Choices for Ancient Woodland*, Peterborough: English Nature.

Evans, J. (1988) *Natural Regeneration of Broadleaves*, Forestry Commisssion Bulletin 78, London: HMSO.

Evans, J. (1992) Coppice forestry: an overview. In G.P. Buckley (ed.) *Ecology and Management of Coppice Woodland*, London: Chapman and Hall.

Evans, J. (1997) Silviculture of hardwoods in Great Britain. *Forestry*, 70 (4): 309–314.

Evans, M.N. and Barkham, J.P. (1992) Coppicing and natural disturbance in temperate woodlands, a review. In G.P. Buckley (ed.) *Ecology and Management of Coppice Woodland*, London: Chapman and Hall.

Feest, A. (1996) Slime moulds: stranger than fiction. *British Wildlife*, 7 (4): 236–242.

Ferris-Kaan, R. and Patterson, G.S. (1992) *Monitoring Vegetation Changes in Conservation Management of Forests*, Forestry Commission Bulletin 108, London: HMSO.

Forest Authority (1992) *Forest Recreation Guidelines*, London: HMSO.

Forestry Commission (1985) *Guidelines for the Management of Broadleaved Woodland*, London: HMSO.

Forestry Commission (1990) *Forest Nature Conservation Guidelines*, London: HMSO.

Forestry Commission (1991a) *Community Woodland Design*, London: HMSO.

Forestry Commission (1991b) *Forests and Water*, London: HMSO.

Forestry Commission (1997) *Forest Life*, No. 13. Edinburgh: Forestry Commission.

Forestry Commission/Countryside Commission (1996) *Woodland Creation: Needs and Opportunities in the English Countryside: A Discussion Paper*, Forestry Commission/Countryside Commission.

Forestry Commission/Countryside Commission (1997) *Woodland Creation: Needs and Opportunities in the English Countryside: Responses to a Discussion Paper*, Forestry Commission/Countryside Commission.

Forest Enterprise (1997) *Corporate Plan 1997–2000*, Edinburgh: Forest Enterprise.

Fry, R. and Lonsdale, D. (1991) *Habitat Choices for Insects: A Neglected Green Issue*, Middlesex: The Amateur Entomologists' Society.

Fuller, R.J. (1982) *Bird Habitats in Britain*, London: Poyser.

Fuller, R.J. (1988) A comparison of breeding bird

assemblages in two Buckinghamshire clay vale woods with different histories of management. In K.J. Kirby and F.J. Wright (eds) *Woodland Conservation and Research in the Clay Vale of Oxfordshire and Buckinghamshire*, NCC Research and Survey in Nature Conservation no. 15, Peterborough: NCC.

Fuller, R.J. (1992) Effects of coppice management on woodland breeding birds. In G.P. Buckley (ed.) *Ecology and Management of Coppice Woodland*, London: Chapman and Hall.

Fuller, R.J. (1995) *Birdlife of Woodland and Forest*, Cambridge: CUP.

Fuller, R.J. and Warren, M.S. (1993) *Coppiced Woodlands: Their Management for Wildlife*, Peterborough: JNCC.

Gibbs, J.N. and Greig, B.J.W. (1997) Biotic and abiotic factors affecting the dying back of pedunculate oak *Quercus robur* L. *Forestry*, 70 (4): 399–406.

Gibson, C.W.D. (1988) The distribution of 'ancient woodland' plants species among areas of different history in Wytham Woods, Oxfordshire. In K.J. Kirby and F.J. Wright (eds) *Woodland Conservation and Research in the Clay Vale of Oxfordshire and Buckinghamshire*, NCC Research and Survey in Nature Conservation no. 15. Peterborough: NCC.

Gilbert, O.L. (1997) Lichen conservation in Britain. In M.R.D. Seaward (ed.) *Lichen Ecology*, London: Academic Press.

Giles, A. (1996) Letter in *Tree News*, Spring: 19.

Godwin, H. (1975) *History of the British Flora: A Factual Basis for Phytogeography*, 2nd edition. Cambridge: CUP.

Greatorex-Davies, J.N. and Marrs, R.H. (1992) The quality of coppice woods and habitats for invertebrates. In G.P. Buckley (ed.) *Ecology and Management of Coppice Woodland*, London: Chapman and Hall.

Green, E.E. (1994) Woodman and the working tree. *Arboricultural Journal*, 88: 205–207.

Grime, J.P., Hodgson, J.G. and Hunt, R. (1988) *Comparative Plant Ecology*, London: Unwin Hyman.

Gurnell, J., Hicks, M. and Whitbread, S. (1992) The effects of coppice management on small mammal populations. In G.P. Buckley (ed.) *Ecology and Management of Coppice Woodland*, London: Chapman and Hall.

Hall, M.L. (1988) Management guidelines for invertebrates, especially butterflies, in plantation woodland. In K.J. Kirby and F.J. Wright (eds) *Woodland Conservation and Research in the Clay Vale of Oxfordshire and Buckinghamshire*, NCC Research and Survey in Nature Conservation no. 15. Peterborough: NCC.

Hambler, C. and Speight, M.R. (1995a) Biodiversity conservation in Britain: Science replacing tradition. *British Wildlife*, 6: 137–147.

Hambler, C. and Speight, M.R. (1995b) Seeing the wood for the trees. *Tree News*, Autumn 1995: 8–11.

Hamilton, G.J. (1985) *Forest Mensuration Handbook*, Forestry Commission Booklet No. 39, London: HMSO.

Hampshire County Council (1991) *Hazel Coppice*, Hampshire County Council.

Harding, P.T. and Alexander, K.N.A. (1993) The saproxylic invertebrates of historic parklands: progress and problems. In K.J. Kirby and C.M. Drake (eds) *Dead wood matters*, Peterborough: English Nature Science No. 7, 58–73.

Harding, P.T. and Rose, F. (1986) *Pasturewoodlands in Lowland Britain*, Huntingdon: ITE.

Harmer, R., Kerr, G. and Boswell, R. (1997) Characteristics of lowland broadleaved woodland being restocked by natural regeneration. *Forestry*, 70 (3): 199–210.

Hart, C. (1991) *Practical Forestry for the Agent and Surveyor*, 3rd edition, Avon: Alan Sutton.

Heliövaara, K. and Väisänen, R. (1984) Effects of modern forestry on northwestern European forest invertebrates: a synthesis. *Acta Forestalia Fennica*, 189: 1–32.

Helliwell, R. (1995) Continuous cover forestry. *Tree News*, Spring/Summer 1995: 12–13.

Hibberd, B.G. (1991) *Forestry Practice*, Forestry Commission Handbook 6, London: HMSO.

Hill M.O., Preston C.D. and Smith A.J.E. (1991–94) *Atlas of the Bryophytes of Britain and Ireland*, Vols 1–3, Colchester: Harley Books.

Hilton, G.M. and Packham, J.R. (1997) A sixteen year record of regional and temporal variation in the fruiting of beech (*Fagus sylvaticus* L.) in England (1980–1995). *Forestry*, 70 (1): 7–15.

HMSO (1988) *The Effects of the Great Storm*, London: HMSO.

Holmes, M. (1997) Bats and trees. *Tree News*, Autumn 1997: 16–17.

Hopkin, S.P. (1997) *Biology of the Springtails*, Oxford: Oxford University Press.

Hornby, R.J. (1988) Woodland conservation and management strategies. In K.J. Kirby and F.J. Wright (eds) *Woodland Conservation and Research in the Clay Vale of Oxfordshire and Buckinghamshire*, NCC Research and Survey in Nature Conservation no. 15, Peterborough: NCC.

Innes, J.L. (1987) *Air Pollution and Forestry*, Forestry Commission Bulletin 70, London: HMSO.

Innes, J.L. and Boswell, R.C. (1987a) *Forest Health Surveys 1987. Part 1. Results*, London: HMSO.

Innes, J.L. and Boswell, R.C. (1987b) *Forest Health Surveys 1987. Part 2. Analysis and Interpretation*, London: HMSO.

Jones, A. (1991) British Wildlife and the Law: a review of the species protection provisions of the Wildlife and Countryside Act 1981. *British Wildlife* 2 (6): 345–358.

Katsaros, P. (1989) *Illustrated Guide to Common Slime Moulds*, California: Mad River Press.

Kennedy, C.E.J. and Southwood, T.R.E. (1984) The number of species of insect associated with British trees: a re-analysis. *Journal of Animal Ecology*, 53: 455–478.

Kirby, K.J. (1984) *Forestry Operations and Broadleaved Woodland Conservation*, Focus on nature conservation no. 8, Peterborough: NCC.

Kirby, K. (1988a) Conservation in British woodlands: adapting traditional management to modern needs. In H.H. Birks, H.J.B. Birks, P.E. Kaland and D. Moe (eds) *The Cultural Landscape: Past, Present and Future*, Cambridge: CUP.

Kirby, K.J. (1988b) *A Woodland Survey Handbook*, Research and Survey in Nature Conservation no. 11, Peterborough: NCC.

Kirby, K.J. (1992) Accumulation of dead wood: a missing ingredient in coppicing? In G.B. Buckley (ed.) *Ecology and Management of Coppice Woodland*, London: Chapman and Hall.

Kirby, K. (1995) Letter in reply to Hambler and Speight 1995a. *British Wildlife*, 6 (5): 338.

Kirby, K. J. and Heap, J. R. (1984) Forestry and Nature Conservation in Romania. *Quarterly Journal of Forestry*, 78: 145–155.

Kirby, K., Mitchell, F.J. and Hester, A.J. (1994) A role for large herbivores (deer and domestic stock) in nature conservation management in British semi-natural woods. *Arboricultural Journal*, 18: 381–399.

Kirby, K.J., Peterken, G.F., Spencer, J.W. and Walker, G.J. (1984) *Inventories of Ancient Semi-Natural Woodland*, Focus on Nature Conservation no. 6, Peterborough: NCC.

Kirby, K.J., Reid, C.M., Thomas, R.C. and Goldsmith, F.B. (1998) Preliminary estimates of fallen deadwood and standing dead trees in managed and unmanaged forests in Britain. *Journal of Applied Ecology*, 35: 148–155.

Kirby, K.J., Webster, S.D. and Anctzak, A. (1991) Effects of forest management on stand structure and the quality of fallen dead wood: Some British and Polish examples. *Forest Ecology and Management*, 43: 167–174.

Kirby, P. (1992) *Habitat Management for Invertebrates: A Practical Handbook*, Peterborough: JNCC/RSPB.

Kirby, P. (1997) Canopy invertebrates (Letter in reply to an article by C. Ozanne). *Tree News*, Autumn 1997: 7.

Kloet, G.S. and Hincks, W.D. (1964–75) *A Checklist of British Insects*, 2nd edition, parts 1–5, London: The Royal Entomological Society.

Kozlowski, T.T., Kramer, P.J. and Pallardy, S.G. (1991) *The Physiological Ecology of Woody Plants*, San Diego: Academic Press Inc.

Lane, A. and Tait, J. (1990) *Practical Conservation: Woodlands*, Milton Keynes: Open University.

Le Sueur, A.D.C. (1931) Burnham Beeches, a study of pollards. *Quarterly Journal of Forestry*, 1–25.

Ling, K.A. and Ashmore, M.R. (1987) *Acid Rain and Trees*, Focus on Nature Conservation no. 19, Peterborough: NCC.

Locket, G.H. and Millidge, A.F. (1951) *British Spiders*, Vols I and II, London: Ray Society.

Lucas, O.W.R. (1997) Aesthetic considerations in British Forestry. *Forestry* 70 (4): 343–349.

Macpherson, G. (1998) Coppice for energy. *Tree News*, Spring 1998: 16–17.

Malcolm, D.C. (1997) The silviculture of conifers in Great Britain. *Forestry*, 70 (4): 293–307.

Margulis L. and Schwartz K. V. (1988) *Five Kingdoms*, 2nd edition, New York: Freeman.

Marren, P. (1990) *Woodland Heritage*, Newton Abbot: David and Charles.

Marren, P. (1992) *The Wildwoods*, Newton Abbot: David and Charles.

Marshall, J.A. and Haes, E.C.M. (1988) *Grasshoppers and Allied Insects of Great Britain and Ireland*, Colchester: Harley Books.

Martin, M.H. and Martin, R.M. (1995) The effect of coppicing on the vegetation of the field and ground layers of an ancient woodland, Lower Wetmoor Wood, *Proceedings of the Bristol Naturalists' Society* 53: 73–84.

Matthews, J.D. (1955) The influence of weather on the frequency of beech mast years in England. *Forestry*, 28: 107–116.

Matthews, J.D. (1994) Implementing forest policy in the lowlands of Britain. *Forestry*, 67 (1): 1–12.

Mitchell, A. (1974) *Field Guide to Trees of Britain and Northern Europe*, London: Collins.

Mitchell, F.J.G. and Kirby, K.J. (1990) The impact of large herbivores on the conservation of semi-natural woodlands in the British uplands. *Forestry*, 63: 333–353.

Mountford, E.P. (1997) A decade of squirrel bark stripping damage to beech in Lady Park Wood U.K. *Forestry*, 70 (1): 17–29.

Myers, N. (1996) The world's forests: problems and potentials. *Environmental Conservation*, 23 (2): 156–168.

NCC (1982) *The Conservation of Semi-natural Upland Woodland*, Peterborough: NCC.

Nichols, D. (1990) *Safety in Biological Fieldwork – Guidance Notes for Codes of Practice*, London: Institute of Biology.

Nilsson, S.G. and Wastljung, U. (1987) Seed predation and cross pollination in mast seeding beech (*Fagus sylvatica*). *Ecology*, 68: 260–265.

Packham, J.R., Harding, D.J.L., Hilton, G.M. and Stuttard, R.A. (1992) *Functional Ecology of Woodlands and Forests*, London: Chapman and Hall.

Perrins, C.M. (1988) Tit populations and beech mast. In K.J. Kirby and F.J. Wright (eds) *Woodland Conservation and Research in the Clay Vale of Oxfordshire and Buckinghamshire*, NCC Research and Survey in Nature Conservation no. 15, Peterborough: NCC.

Peterken, G.F. (1992) Coppices in the lowland landscape. In G.P. Buckley (ed.) *Ecology and Management of Coppice Woodland*, London: Chapman and Hall.

Peterken, G.F. (1993) *Woodland Conservation and Management*, 2nd edition. London: Chapman and Hall.

Peterken, G.F. (1996) *Natural Woodland*, Cambridge: CUP.

Peterken, G.F. and Backmeroff, C. (1988) *Long-Term Monitoring in Unmanaged Woodland Nature Reserves*, Research and survey in nature conservation no. 9, Peterborough: NCC.

Peterken, G.F. and Hughes, F.M.R. (1995) Restoration of floodplain forests in Britain. *Forestry*, 68 (3): 187–202.

Peterken, G.F. and Mountford, E.P. (1996) Effects of drought on beech in Lady Park Wood, an unmanaged mixed deciduous woodland. *Forestry*, 69 (2): 125–136.

Peterson, R., Mountfort, G. and Hollom, P.A.D. (1983) *A Field Guide to the Birds of Britain and Europe*, 4th edition, London: Collins.

Phillips, R. (1981) *Mushrooms*, London: Pan.

Pilcher, N. (1996) Letter in reply to Hambler and Speight. *Tree News*, Spring 1996: 18.

Power, S.A. (1994) Temporal trends in twig growth of *Fagus sylvatica* L. and their relationships with environmental factors. *Forestry*, 67 (1): 13–300.

Rackham, O. (1975) *Hayley Wood: Its History and Ecology*. Cambridge: Cambs and Isle of Ely Naturalists' Trust.

Rackham, O. (1980) *Ancient woodland: Its History, Vegetation and Uses in England*, London: E. Arnold.

Rackham, O. (1988) Trees and woodland in a crowded landscape: the cultural landscape of the British Isles. In H.H. Birks, H.J.B. Birks, P.E. Kaland and D. Moe (eds) *The Cultural Landscape: Past, Present and Future*, Cambridge: CUP.

Rackham, O. (1990) *Trees and Woodland in the British Landscape*, London: J.M. Dent.

Rackham, O. (1997) Where is Beech native? *Tree News*, Autumn 1997: 8–9.

Ratcliffe, P. (1992) The interaction of deer and vegetation in coppiced woods. In G.P. Buckley (ed.) *Ecology and Management of Coppice Woodland*, London: Chapman and Hall.

Read, H.J. and Frater, M. (1993) Maximising woodland conservation value through management. In M.E.A. Broekmeyer, W. Vos and H. Koop (eds) *European Forest Reserves*, Wageningen: Pudoc Scientific Publishers.

Read, H.J., Frater, M. and Noble, D. (1996) A survey of the condition of the pollards at Burnham Beeches and results of some experiments in cutting them. In H.J. Read (ed.) *Pollard and Veteran*

Tree Management II, London: Corporation of London.

Read, H.J., Frater, M. and Turney, I.S. (1991) Pollarding in Burnham Beeches, Bucks: a historical review and notes on recent work. In H.J. Read (ed.) *Pollard and Veteran Tree Management*, London: Corporation of London, pp. 11–18.

Redfern, M. and Askew, R.R. (1992) *Plant Galls*, Naturalists' Handbooks 17, Slough: Richmond Publishing Co. Ltd.

Richardson, D.H.S. (1981) *Pollution Monitoring with Lichens*, Naturalists' Handbooks 19, Slough: Richmond Publishing Co. Ltd.

Roberts, M.J. (1985) *The Spiders of Britain and Ireland*, vol. 1, Colchester: Harley Books.

Roberts, M.J. (1987) *The Spiders of Britain and Ireland*, vol. 2, Colchester: Harley Books.

Roberts, M.J. (1995) *Spiders of Britain and Northern Europe*, London: HarperCollins.

Robinson, R.K. (1972) The production by roots of *Calluna vulgaris* of a factor inhibitory to growth of some mycorrhizal fungi. *Journal of Ecology*, 60: 219–224.

Rodwell, J. (ed.) (1991) *British Plant Communities. Vol. 1 Woodlands and Scrub*, Cambridge: Cambridge University Press.

Rodwell, J. and Patterson, G. (1994) *Creating New Native Woodlands*, Forestry Authority Bulletin 112, London: HMSO.

Rollinson, T.J.D. and Evans, J. (1987) *The Yield of Sweet Chestnut Coppice*, Forestry Commission Bulletin 64, London: HMSO.

RSPB (1993) *Time for Pine*, Edinburgh: RSPB.

Schofield, P. (1988) Final discussion and summing up. In K.J. Kirby and F.J. Wright (eds) *Woodland Conservation and Research in the Clay Vale of Oxfordshire and Buckinghamshire*, NCC Research and Survey in Nature Conservation no. 15, Peterborough: NCC.

Sheail, J. Treweek, J.R. and Mountford, J.O. (1997) The U.K. transition from nature preservation to 'creative conservation'. *Environmental Conservation*, 24 (3): 224–235.

Shirt, D.B. (ed.) (1987) *British Red Data Books: 2. Insects*, Peterborough: Nature Conservancy Council.

Skinner, B. (1984) *Moths of the British Isles*. London, Viking.

Smart, N. and Andrews, J. (1985) *Birds and Broadleaves Handbook*, Bedfordshire: RSPB.

SNH (n.d.) *Ancient Woodland in Scotland*, Edinburgh: SNH.

Soutar, R. and Peterken, G. (1989) Native trees and shrubs for wildlife. *Tree News*, September 1989.

Stace C. (1997) *New Flora of the British Isles*, 2nd edition, Cambridge: Cambridge University Press.

Steane, J.M. (1988) Timber exploitation in central-southern England during the 13th–15th centuries. In K.J. Kirby and F.J. Wright (eds) *Woodland Conservation and Research in the Clay Vale of Oxfordshire and Buckinghamshire*, NCC Research and Survey in Nature Conservation no. 15, Peterborough: NCC.

Steel, C. (1988) Butterfly monitoring in Sheephouse Wood. In K.J. Kirby and F.J. Wright (eds) *Woodland Conservation and Research in the Clay Vale of Oxfordshire and Buckinghamshire*, NCC Research and Survey in Nature Conservation no. 15, Peterborough: NCC.

Steel, D. and Mills, N. (1988) A study of plants and invertebrates in an actively coppiced woodland (Brasenose Wood, Oxfordshire). In K.J. Kirby and F.J. Wright (eds) *Woodland Conservation and Research in the Clay Vale of Oxfordshire and Buckinghamshire*, NCC Research and Survey in Nature Conservation no. 15, Peterborough: NCC.

Stephenson, S. and Stempen, H. (1994) *Myxomycetes: A Handbook of Slime Moulds*, Portland, OR: Timber Press.

Sterling, P.H. and Hambler, C. (1988) Coppicing for conservation: Do hazel communities benefit? In K.J. Kirby and F.J. Wright (eds) *Woodland Conservation and Research in the Clay Vale of Oxfordshire and Buckinghamshire*, NCC Research and Survey in Nature Conservation no. 15, Peterborough: NCC.

Stubbs, A. (1997) Removal of fallen timber from the River Wye and its tributaries in England and Wales. *Invertebrate Conservation News*, 24: 8–10.

Stubbs, A. and Chandler, P. (1978). *A Dipterist's Handbook*. Middlesex: The Amateur Entomologists' Society.

Tansley, A.G. (ed.) (1911) *Types of British Vegetation*, Cambridge: Cambridge University Press.

Tansley, A.G. (1939) *The British Islands and their Vegetation*, Cambridge: CUP.

Thomas, R.C., Kirby, K.J. and Reid, C.M. (1997)

The conservation of a fragmented ecosystem within a cultural landscape. The case for ancient woodland in England. *Biological Conservation*, 82: 243–252.

Thomas, J. and Lewington, R. (1991) *The Butterflies of Britain and Ireland*, London: Dorling Kindersley.

Thurkettle, V. (1997) The marketing of British hardwoods. *Forestry*, 70 (4): 319–325.

Tittensor, A. (1980) *The Red Squirrel*, Poole: Blandford Press.

Tubbs, C.R. (1986) *The New Forest*, London: Collins (New Naturalist).

Tubbs, C.R. (1997) The ecology of pastoralism in the New Forest. *British Wildlife*, 9 (1): 7–16.

Waring, P. (1988) Responses of moth populations to coppicing and the planting of conifers. In K.J. Kirby and F.J. Wright (eds) *Woodland Conservation and Research in the Clay Vale of Oxfordshire and Buckinghamshire*. NCC Research and Survey in Nature Conservation no. 15, Peterborough: NCC.

Wareing, P.F. and Phillips, I.D.J. (1981) *Growth and Differentiation in Plants*, 3rd edition, Oxford: Pergamon Press.

Warren, M.S. and Key, R.S. (1991) Woodlands: past, present and potential for insects. In M. Collins and J.A. Thomas (eds) *The Conservation of Insects and Their Habitats: Proceedings of the 15th Symposium of the Royal Entomological Society of London, 1989*, London: Academic Press, pp. 155–211.

Warren, M.S. and Thomas, J.A. (1992) Butterfly responses to coppicing. In G.P. Buckley (ed.) *Ecology and Management of Coppice Woodland*, London: Chapman and Hall.

Watkins, C. (1990) *Woodland Management and Conservation*, Newton Abbot: David and Charles.

Watson, E.V. (1981) *British Mosses and Liverworts*, 3rd edition, Cambridge: CUP.

Welch, R.C. (1969) Coppicing and its effect on woodland invertebrates. Devon Trust for Nature Conservation, 22: 969–973.

Whitbread, A.M and Kirby, K.J. (1992) *Summary of National Vegetation Classification Woodland Descriptions*, U.K. Nature Conservation No. 4, Peterborough: JNCC.

White, J. (1995) *Forest and Woodland Trees in Britain*, Oxford: OUP.

Worrell, R. and Nixon, C.J. (1991) *Factors Affecting the Natural Regeneration of Oak in Upland Britain*, Forestry Commission occasional Paper No. 31, Edinburgh.

Zuidema, P.A., Sayer, J.A. and Dijkman, W. (1996) Forest fragmentation and biodiversity: the case for intermediate-sized conservation areas. *Environmental Conservation*, 23 (4): 290–297.

SUBJECT INDEX

•

SPECIES INDEX

•